W9-DAY-523

SPACE AND SECURITY

Selected Titles in ABC-CLIO's
CONTEMPORARY
WORLD ISSUES
Series

For a complete list of titles in this series, please visit
www.abc-clio.com

Books in the Contemporary World Issues series address vital issues in today's society, such as genetic engineering, pollution, and biodiversity. Written by professional writers, scholars, and nonacademic experts, these books are authoritative, clearly written, up-to-date, and objective. They provide a good starting point for research by high school and college students, scholars, and general readers as well as by legislators, businesspeople, activists, and others.

Each book, carefully organized and easy to use, contains an overview of the subject, a detailed chronology, biographical sketches, facts and data and/or documents and other primary source material, a directory of organizations and agencies, annotated lists of print and nonprint resources, and an index.

Readers of books in the Contemporary World Issues series will find the information they need to have a better understanding of the social, political, environmental, and economic issues facing the world today.

SPACE AND SECURITY

A Reference Handbook

Peter L. Hays

**CONTEMPORARY
WORLD ISSUES**

 ABC-CLIO

Santa Barbara, California • Denver, Colorado • Oxford, England

HoT TopIC
629.4
H425
2011

Copyright 2011 by ABC-CLIO, LLC

Library of Congress Cataloging-in-Publication Data

Hays, Peter L.
   Space and security : a reference handbook / Peter L. Hays.
      p. cm. — (Contemporary world issues series)
   Includes bibliographical references and index.
   ISBN 978-1-59884-421-4 (alk. paper) — ISBN 978-1-59884-422-1 (ebook)
1. Astronautics and state—United States. 2. Astronautics, Military—United States. 3. Space industrialization—Government policy—United States. 4. Space surveillance—Government policy—United States. 5. Outer space—Civilian use—Government policy—United States. I. Title.
   UG1523.H395   2011
   629.40973—dc22   2010041454

ISBN: 978-1-59884-421-4
EISBN: 978-1-59884-422-1

15 14 13 12 11   1 2 3 4 5

This book is also available on the World Wide Web as an eBook.
Visit www.abc-clio.com for details.

ABC-CLIO, LLC
130 Cremona Drive, P.O. Box 1911
Santa Barbara, California 93116-1911

This book is printed on acid-free paper ∞
Manufactured in the United States of America

Contents

List of Tables

Preface

This book provides a comprehensive review of how the United States has developed and implemented policies to use space capabilities to enhance national security from before *Sputnik* to the present, focusing, in particular, on areas of continuity and change. Chapter 1 presents the foundational background and history of U.S. national security space policy. It emphasizes how the United States sought to legitimize satellite overflight for reconnaissance purposes in order to open up the closed Soviet state and then organized to manage the national security space enterprise throughout the remainder of the Cold War. This foundation is essential for understanding current national security space-policy issues. Chapter 2 examines key issues that present opportunities and challenges for U.S. national security space policy including military transformation, export controls, and missile defense. Moving to a global perspective, Chapter 3 discusses space law, space weaponization, increasing congestion, and the emergence of China as a major space power. How the United States and others address these issues will be critical in shaping space security and determining whether it can evolve in transparent, stable, and sustainable ways. Chapter 4 is a chronology covering key events in the development and implementation of U.S. national security space policy from 1954 to 2010. Biographical sketches of all key actors in U.S. national security space policy are in Chapter 5, including all U.S. Presidents from Eisenhower to Obama, important military figures, and key civilian decision makers. Chapter 6 contains excerpts from the most relevant statements about U.S. national security space policy from 1955 to the present, providing rich detail and specific context about these policies. Chapter 7 is a directory of all organizations that played a significant role in the development and implementation of U.S. national security space policy. Organizations are listed chronologically, annotated with brief descriptions of their relevance to national security space policy, and online contact information is provided. Finally, Chapter 8 includes a comprehensive listing of printed and online resources related to national security space issues.

I would like to thank my colleagues in the national security space enterprise, my colleagues and students at the Space Policy Institute at George Washington University, and particularly my family for all their inspiration and support of this effort. My son, Peter Dillon, compiled and organized the material for the second half of this book and deserves special recognition. Finally, the editorial staff at ABC-CLIO masterfully orchestrated all the work needed to turn the manuscript into a textbook. I owe all these individuals a debt of gratitude but, of course, any remaining errors are solely my responsibility.

Peter L. Hays
Fairfax Station, Virginia
August 2010

1

Background and History

Humans have pondered their relationship with the cosmos for millennia, but only during the past several generations have we begun moving beyond just dreaming by developing policies and building some of the tools needed to "leave the cradle." Individuals such as Jules Verne, H. G. Wells, Konstantin Tsiolkovsky, Robert Goddard, Wernher von Braun, Eugen Sanger, and Sergei Korolev helped to lay the theoretical and practical foundations of spaceflight. The Soviet Union's opening of the space age with the launch of *Sputnik* I on October 4, 1957, was a first step toward realizing the dreams and visions of these and countless others, forever altering humanity's perception of itself and its place in the cosmos. Our visions of space and of our role and purpose in the cosmos are as varied as individuals themselves and include ideas as sweeping, extreme, and contradictory as universal peace and total domination. At the opening of the space age it was unclear what form these visions would take and today it remains uncertain how humanity's entry into space will ultimately affect life on Earth and beyond.

This book describes and explains the policies that have guided United States national-security efforts in space from before *Sputnik* to the present. Because it can easily be more than 30 years from the time space systems are first proposed to the end of their operational life, this chapter focuses more attention on the evolution of space policy through the end of the Cold War; current domestic and international space-policy issues and challenges are addressed in the following two chapters. This chapter also explains how U.S. policies have fostered development and evolution of space activity in four interconnected sectors: civil, commercial, military, and intelligence. Civil space activity seeks to explore and understand space, focusing on how space can improve and protect life on Earth. The International Space Station is the centerpiece of current civil space activity and future efforts may include continuing human exploration and potential exploitation of the Moon, flexible path missions to asteroids and other relatively close destinations, and human exploration of Mars. Commercial space activity is designed to generate wealth from and in space and has grown to become an increasingly important and competitive part of the global economy. Defense space sector capabilities enable U.S. leaders to implement American foreign policy and, when necessary, employ increasingly potent and precise military power in ways never before possible. Considerable controversy surrounds U.S. policies to develop capabilities to protect its space capabilities against hostile acts and, especially, to counter hostile use of space against U.S. interests. Finally, within the intelligence space sector, the United States collects a wide range of data such as high resolution imagery that is essential to the formulation and execution of foreign and defense policies.

U.S. Space Policy before *Sputnik*

A desire for better intelligence on the strategic capabilities and intentions of the closed Soviet state became a primary security concern for the United States at the onset of the Cold War. Several ominous developments in the late 1940s and early 1950s underscored this need including inaccurate predictions on when the Soviets would first develop atomic weapons and have operational thermonuclear weapons, uncertainties about a possible bomber gap, the failure of President Dwight Eisenhower's "Open Skies" proposal of July 1955, and many issues related to the progress and strategic impact of the Soviet intercontinental ballistic missile (ICBM) program.

Even in the earliest days of the Cold War some visionaries believed that space might provide an ideal vantage point for spying on the Soviets. The RAND Corporation was created in March 1946 and its first report, "Preliminary Design of an Experimental World-Circling Spaceship," completed in April 1946, not only explained space physics and spaceship technical designs but also introduced almost every one of the major military space missions that would be developed in the coming years including communications, attack assessment, weather reconnaissance, and strategic reconnaissance. RAND also became the first organization to comprehensively analyze the political implications of the opening of the space age in an October 1950 report that highlighted the likely psychological impression the first satellite would leave on the public and raised the critical political issues of "overflight" and "freedom of space"—asking how the Soviets would respond to new issues in international law such as satellites flying over and photographing their territory. The report suggested that one way to test the issue of freedom of space would be first to launch an experimental U.S. satellite in an equatorial orbit that would not cross Soviet territory before attempting any satellite reconnaissance overhead the Soviet Union.

By the mid-1950s, the development of aircraft and photoreconnaissance satellites with the potential to help open the closed Soviet state had become the top U.S. space-policy goal. To support this highest priority objective, if aerial reconnaissance flights could not be implemented, U.S. space policy concurrently sought to build and protect a legal regime designed to legitimize the operation of spy satellites. President Eisenhower's approach toward space and strategic issues was strongly influenced by the top secret Technological Capabilities Panel (TCP) he commissioned in March 1954. Eisenhower chose James Killian, president of the Massachusetts Institute of Technology (MIT), as chairman of the TCP and made it clear that he wanted the best minds in the country to focus on the technological problem of preventing another Pearl Harbor.

As part of the TCP process, Killian and Edwin (Din) Land, founder of the Polaroid Corporation and chairman of the intelligence subcommittee of the TCP, were briefed on a wide range of potential collection methods and systems including satellites, but they became most enthused about attempting high-altitude reconnaissance overflights of the Soviet Union via a jet-powered glider they saw on the drawing boards at Clarence (Kelly) Johnson's Lockheed "Skunk Works" in Burbank, California. They recommended production of this new aircraft during a series of briefings culminating in an Oval Office meeting on November 24, 1954, attended by the President, Secretaries of State and Defense, as well as top Department of Defense (DoD) and Central Intelligence Agency (CIA) officials. The initial programs and structure for a top-priority national strategic reconnaissance program were discussed at this meeting and the President verbally authorized the CIA to begin developing the U-2 aircraft with Air Force support, later adding the proviso that CIA pilots operate the planes.

The TCP completed a secret two-volume report and briefed the National Security Council (NSC) in February 1955. The report strongly recommended rapid development of U.S. technical intelligence-gathering capabilities and supporting policies for overflight. Land wrote:

> We *must* find ways to increase the number of hard facts upon which our intelligence estimates are based, to provide better strategic warning, to minimize surprise in the kind of attack, and to reduce the danger of gross overestimation or gross underestimation of the threat. To this end we recommend adoption of a vigorous program for the extensive use, in many intelligence procedures, of the most advanced knowledge in science and technology. (U.S. Dept of State 1990a, 54, emphasis in original)

The report urged construction and launch of a small scientific Earth satellite that would operate above the sovereign airspace of states and emphasized that "although it is clear that a very small satellite cannot [now] serve as useful carrier for reconnaissance apparatus . . . it can serve ideally to explore or establish the principle that space, outside our atmosphere, is open to all." A small satellite might help establish the principle of freedom of space in international law and create a useful precedent when "we probably shall conduct extensive reconnaissance by means of large and complex satellites"(Technological Capabilities 1955, 147–148). The TCP process and report were critical drivers behind development of America's first high-tech intelligence collection platforms: the Lockheed U-2 aircraft and the weapons system (WS)-117L reconnaissance satellites.

Based on the TCP report and presidential direction, the Air Force initiated America's first space program in November 1954 and set formal but secret requirements including three axis stabilization and high-pointing accuracy for the WS-117L satellite program on March 16, 1955. Brigadier General Bernard Schriever had taken command of the newly established Western Development Division (WDD) in El Segundo, California, in August 1954 and was enthusiastic about developing WS-117L satellites but was primarily charged with developing the higher-priority Atlas ICBM and received little funding or support for developing satellites. Nonetheless, even prior to the funding increases triggered by *Sputnik*, Air Force work on the WS-117L evolved to encompass secret research efforts on each of the four primary types of reconnaissance and surveillance satellite systems that would be used over the next 30 years: reconnaissance via recoverable film systems that became the Corona program, reconnaissance via electro-optical systems under the Samos program, infrared surveillance for missile launch detection in the Midas program, and satellites to detect nuclear detonations under the Vela Hotel program.

U.S. space policies designed to support the exploration of space for scientific purposes proceeded along a parallel track during this period and are an important part of the development of early U.S. space policy because, while they were lower priority than the top secret U.S. spy-satellite development efforts, they were the highest priority open U.S. space efforts and affected how both the U.S. government and public considered uses of space. In addition, the pace and structure of U.S. space-science efforts played a crucial role in the timing of the first U.S. satellites. Considerable scientific interest within the United States and around the world in developing satellites to explore the upper atmosphere became the impetus behind creation of an International Geophysical Year (IGY) to be held between July 1, 1957, and December 31, 1958. The idea for an IGY focused on high-altitude research was first discussed at an informal meeting of a group of American space scientists near Washington in April 1950 and by March 1955 the presidents of the National Academy of Science and the National Science Foundation had met with Eisenhower and received his support. In July 1955, White House press secretary James Hagerty publicly announced "the President has approved plans by this country for going ahead with the launching of small, Earth-circling satellites as part of the United States participation the International Geophysical Year" (Quoted in McDougall 1985, 118–119).

The TCP report and the IGY satellite proposal were major inputs that prompted the NSC to undertake a delicate and hidden task—development of America's first national space policy. This effort focused on prioritizing and harmonizing the inchoate space goals and programs of the United States during the mid-1950s but ended up laying a solid foundation for all subsequent U.S. space policy. During the spring of 1955, the NSC Planning Board, Special Assistant to the President on Government Operations Nelson Rockefeller, and Assistant Secretary of Defense for Research & Development Donald Quarles reviewed and analyzed the differing goals and requirements of the TCP "freedom of space" objective, the WS-117L program, the U.S. IGY satellite proposal, and several military requirements and booster considerations. On May 20, Quarles submitted to the NSC a draft that became the basis for the secret document labeled NSC 5520, "Draft Statement of Policy on U.S. Scientific Satellite Program," that was approved by President Eisenhower on May 27.

NSC 5520 noted that the Soviets were hard at work with their own IGY satellite efforts and recognized that "considerable prestige and psychological benefits will accrue to the nation which first is successful in launching a satellite." It emphasized that "a small scientific satellite will provide a test of the principle of freedom of space." Accordingly, the report recommended the United States launch a U.S. scientific satellite program during the IGY and recognized that this effort represented "an excellent opportunity" publicly to emphasize and link the United States to the scientific and peaceful purposes of its first projected satellite and more generally characterize U.S. space efforts this way. At the same time, however, the report also emphasized factors considered more important than the IGY program: preserving "U.S. freedom of action in the field of satellites and related programs"; not delaying or otherwise impeding other U.S. satellite programs; protecting U.S. classified information; and in no way "imply[ing] a requirement for prior consent by any nation over which the satellite might pass in its orbit" or "jeopardiz[ing] the concept of Freedom of Space" (U.S. Dept of State 1990b, 725–730). Overall, NSC 5520 meant public support for the U.S. IGY satellite proposal but in secret meant political and programmatic primacy for the WS-117L program and plans to use the benign IGY program as the first test of the Soviet response to the overflight issue. In other words, the IGY satellite was to serve as a "stalking horse" to establish the precedent of space overflight and legitimize eventual operation of military reconnaissance satellites (Hall 1995, 217–218).

Following this secret NSC maneuvering and the public announcement the United States would launch a satellite during the IGY, the remaining major issue was how the IGY satellite would be launched into space. Only military booster technology offered the chance to launch within the rapidly approaching IGY window. Quarles named an advisory group of scientists chaired by Homer Stewart of the Jet Propulsion Laboratory (JPL) to study the booster question. Each of the services made presentations to the Stewart Committee in July 1955 and competed for the honor of having its booster open the space age. The Air Force proposed launching a large IGY satellite atop its top priority Atlas ICBM but could not guarantee this would not interfere with the development of the Atlas or even that the Atlas would be ready in time for the IGY; this proposal was quickly eliminated. Army ballistic missile experts from the Redstone Arsenal including Wernher von Braun offered the Stewart Committee their previously developed (September 1954) Project Orbiter proposal for a small satellite to be launched atop a V-2 derived Redstone booster. The Naval Research Laboratory's (NRL) proposal called for the development of an upgraded version of the Navy's Viking sounding rocket capable of launching a very small satellite. On August 3, 1955, the Stewart Committee voted 3–2 in favor of the NRL proposal. Many factors were at work influencing this close vote including political sensitivities and a desire to avoid having a Nazi-tainted booster lead the United States into the space age. Selection of the unproven NRL booster by the Stewart Committee reemphasized the civilian face of U.S. space efforts, covertly supported overflight considerations, but also set the stage for America's second-place finish in the first space race.

Sputnik Changes Everything, 1957–1963

The Soviet Union became the world's first spacefaring nation with the launch of *Sputnik* I on October 4, 1957. The Soviets rapidly followed this premiere by orbiting the 1,121 pound *Sputnik* II with *Laika* the dog aboard on November 3. This potent one-two punch wreaked havoc on Americans' preferred image as the world's leader in science and technology, deepening insecurities the nuclear age had already left on the American psyche. The U.S. response to the *Sputniks* challenge was both broad and deep, ranging from many new educational approaches and programs to increased military spending and reorganization. The shock of the *Sputniks* was the proximate cause of nearly all the space-related developments in the United States during 1957 and 1958 including Senator Lyndon Johnson's Preparedness Subcommittee hearings, the creation of the Advanced Research Projects Agency (ARPA) within DoD, and the establishment of the National Aeronautics and Space Administration (NASA).

Despite warnings concerning the public impact of the first satellite in NSC 5520 and the many public statements on impending satellite launches for the IGY, neither the Eisenhower Administration nor the American public was well prepared for how the space age began. In retrospect, failure to anticipate and plan for the American public's reaction to being beaten into space was a great weakness in Eisenhower's space policy. The pervasive atmosphere of uncertainty and media hysteria that developed after the *Sputniks* demanded a significant response yet Eisenhower's complex and largely hidden space-policy goals placed his administration in a nearly indefensible public-policy position—a situation that could only be improved via a U.S. satellite success or a self-defeating open discussion of putting a civilian face on secret efforts to build spy satellites and legitimize their operation. On October 8, Eisenhower privately called Quarles to the White House to explain the situation and their discussion was captured in the detailed notes taken by Eisenhower's personal aide, Brigadier General Andrew Goodpaster:

> There was no doubt, Quarles confessed, that the Redstone, had it been used, could have orbited a satellite a year or more ago. Ike said that when this information reached Congress they would surely ask why such action was not taken. The President recalled, however, that timing was never given too much importance in our own program, which was tied to the IGY, and confirmed that, in order for all the scientists to be able to look at the instrument, it had to be kept away from military secrets. Quarles then accentuated the positive saying the Russians have in fact done us a good turn, unintentionally, in establishing the concept of freedom of international space. The President then looked ahead five years, and asked about a reconnaissance vehicle.[1] (McDougall 1985, 134)

At this same meeting Eisenhower agreed to Quarles's request to study using the Jupiter-C as a quicker and surer way to launch America's first satellite. Incoming Secretary of Defense Neil McElroy was visiting the Army Ballistic Missile Agency (ABMA) on October 4 when the news of *Sputnik* broke. Following this news flash, von Braun and ABMA Commander Major General John Medaris cornered McElroy and asserted that ABMA's Project Orbiter could place a satellite in orbit within 90 days. Von Braun and Medaris could give McElroy these assurances because a Jupiter-C launched in September 1956 could have placed a satellite into orbit (it reached 600 miles high and traveled 3,000 miles downrange; it lacked only a fourth stage "kick motor" to place a satellite into orbit) and they had carefully stored two Jupiter-Cs for such a contingency. Following the televised spectacular failure of America's first satellite launch attempt with the December 6 launch-pad explosion of Vanguard TV-3, the Jupiter-C was given the go-ahead and successfully launched *Explorer* I, America's first satellite, from Cape Canaveral on January 31, 1958.

The *Sputnik* shock also underscored the interrelationships between U.S. national security and the state of U.S. science and technology. On October 15, Eisenhower met with the distinguished scientists who made up the Science Advisory Committee and later decided to appoint James Killian to the new position of Special Assistant to the President for Science and Technology. In early November, Eisenhower and the NSC received an important report from a top secret strategic review committee commissioned in the spring of 1957. This review, known as the Gaither Report after committee chairman Rowan Gaither, presented a very somber picture of the current and future strategic balance between the superpowers and was especially concerned with the vulnerability of the U.S. bomber force in the space and missile age. While Eisenhower believed the report was a worst-case analysis and rejected its tone and many of its recommendations, the report highlighted a widening split between the President and the growing number of his detractors over his response to the *Sputniks*, the perceived missile gap, and his general defense policy.

The Air Force had moved most quickly and directly of all the services to claim jurisdictional control of U.S. military operations in outer space. The first comprehensive and official expression of Air Force space doctrine was given by Air Force Chief of Staff General Thomas White in a presentation at the National Press Club on November 29, 1957. White first asserted that just as "whoever has the capability to control the air is in a position to exert control over the lands and seas . . . in the future whoever has the capability to control space will likewise possess the capability to exert control of the surface of the earth." This claim was linked directly to the central but still controversial Air Force doctrinal tenet regarding the necessity for air superiority for success in any other military operation; White further asserted that the United States "must win the capability to control space." White's second major doctrinal assumption addressed the aerospace concept: "there is no division, *per se*, between air and space. Air and space are an indivisible field of operations" (Reprinted in Emme 1959, 496–501). White's presentation built a case for supporting Air Force efforts to control this indivisible medium, bureaucratically tied the service to the aerospace concept, and implied the Air Force was the service best prepared to respond to the grave national security challenges of space; but, of course, placed the Air Force in direct conflict with Eisenhower's policy on space as a sanctuary for the development, use, and protection of reconnaissance satellites.

One of the most important, if not the most important, single factors in shaping American attitudes on space and security issues after *Sputnik* was the Senate hearings called by majority leader Lyndon Johnson. These hearings provided a forum for top civilian and military leaders to express their views about space and U.S. national security, were critical in shaping American's perception of space, helped to channel early American space policy onto certain paths, and even helped to mold the historiography of the opening of the space age. Unfortunately, however, the hearings did not even come close to uncovering the secret Eisenhower space-policy requirements driven by spy satellites and overflight considerations and thus could not begin to encourage informed debate or produce a complete or coherent picture of this complex policy. Most of the military witnesses were united in their opposition to the creation of a new unified military space command or civilian space agency, but on this point von Braun parted company with his military masters—much of his testimony emphasized prospective civil and scientific space tasks he proposed be undertaken by a national space agency. Overall, the hearings are best remembered for presenting a very different and more alarmist viewpoint about space, a position summarized by Johnson's final report: "We are in a race for survival and we intend to win that race" (Divine 1993, 79).

Announcement of the Advanced Research Project Agency (ARPA) by Secretary McElroy on November 15, 1957, was another development that rapidly followed the initial *Sputniks* crisis. Establishing a new space organization within DoD was a way at least temporarily to derail early congressional efforts to advance their own, more wide-ranging solutions to perceived problems with U.S. space policy. A new organization controlled by the Office of the Secretary of Defense (OSD) was also a way to circumvent Air Force bureaucratic efforts to gain greater control over all military space programs. APRA was charged with two main responsibilities: directing all military space programs and initiating programs and actions deemed necessary to avoid another *Sputnik*. In congressional testimony in 1959, Director Roy Johnson was very clear about ARPA's primary purpose: "The Defense Secretary is very concerned about programs where all three services have a common interest, to prevent duplication. He wants one space program, not three" (U.S. Congress 1959, 532). For a time ARPA directed almost all U.S. space efforts; for example, in February 1958 it assumed control of the WS-117L program and this became the agency's single most important space project, accounting for $152 million or nearly one-third of ARPA's 1958 budget. Following DoD reorganizations in 1959 ARPA returned primary responsibility for most military space R&D to the Services.

An additional initial programmatic response to the *Sputniks* came on February 3, 1958, when Eisenhower directed that development of intercontinental and intermediate range ballistic missiles (ICBMs and IRBMs) as well as the WS-117L satellite programs be given highest and equal national priority. Then at a February 7 meeting with Killian, Land, and Goodpaster, the President decided to make the CIA, rather than the Air Force, primarily responsible for the development of a reconnaissance satellite using the recoverable film method. This program was designated Corona and was scheduled to be operational by the spring of 1959.

All the satellite systems being developed under the WS-117L program experienced significant technological difficulties before becoming operational. The Corona program began with the lift-off of *Discoverer* I from Vandenberg Air Force Base (AFB), California, on February 28, 1959, but significant technological reliability problems with the Thor-Agena launch vehicles as well as various control glitches with the satellites themselves and with the film recovery system prevented any successful film recoveries from the 12 launches between February 1959 and June 1960. Meanwhile, the more technologically demanding Samos system was not even ready to begin flight testing and the Midas infrared missile launch detection program was experiencing similar large technological challenges. At the White House, Science Adviser George Kistiakowsky believed that the Air Force was putting too much effort into the direct electro-optical data return Samos program, based on technologies he believed would not mature for some time, and that this overemphasis was disrupting the entire spysat development effort.

In May 1960, Eisenhower directed Kistiakowsky to set up a committee to study what corrective actions might be necessary. Kistiakowsky and Defense Secretary Thomas Gates decided on a study committee composed of three people: Under Secretary of the Air Force Joseph Charyk, Deputy Director of Defense Research and Engineering (DDR&E) John Rubel, and Kistiakowsky. This group, which came to be known as the Samos Panel, reported their recommendations to the NSC on August 25. Their primary recommendation, immediate creation of an organization to provide a direct chain of command from the Secretary of the Air Force to the officers in charge of each spysat project, was enthusiastically supported by Eisenhower and led to the formal establishment in August 1961 of the highly classified National Reconnaissance Office (NRO)—the very existence of which was an official U.S. state secret until September 1992.

Corona became the world's first operational satellite photo-reconnaissance system following the successful retrieval of the film canister ejected from *Discoverer* XIV on August 19, 1960. There were three more successful film retrievals during the remainder of 1960 and the data from these missions had a nearly immediate and profound effect on the U.S. view of the strategic balance as data from these space-based collection systems, together with the more limited data available from U-2 overflights, was able to lay the missile-gap issue to rest. The first National Intelligence Estimate (NIE) to incorporate spysat data was completed in September 1961 and stated that the United States believed the Soviets had fewer than 10 operational ICBMs—a far less threatening force than the 500 Soviet ICBMs predicted for 1961 in the November 1957 NIE. Major lessons learned from the missile-gap controversy highlighted the importance of overhead reconnaissance and pointed to the need to separate intelligence producers from intelligence consumers, reinforcing the decision to create NRO and preclude this problem with the interpretation of spysat data.

Of the many questions facing the administration and Congress as they struggled to craft a civilian space agency during the spring and summer of 1958, none were more important than the issues surrounding the proper relative priority of civil versus military space efforts and the likely bureaucratic impact of a new civilian space agency. The National Aeronautics and Space Act of 1958 that created NASA was the single most important response of the U.S. government to the *Sputniks* challenge and represents a true compromise created out of many conflicting bureaucratic interests and policy goals. In early 1958, following the Johnson hearings and a spate of proposed space and science legislation, Congress was clearly in the mood to consider far more sweeping organizational changes in the way America conducted its space business than had been accomplished through the creation of ARPA. The President's Science Advisory Committee (PSAC) had spent the last months of 1957 in a series of debates over the relative value of various potential space missions and had considered many different ways to organize the government to conduct space activity. By the end of December, a consensus emerged that scientifically oriented civil space missions ought to be the nation's top space priority and that a civilian space agency built from and modeled after the National Advisory Committee on Aeronautics (NACA) would be the best organizational approach.

The administration began developing proposed legislation for a national space agency following key meetings at the White House on the third and fourth of February 1958. On February 3, the PSAC was formally tasked with studying space mission priorities and recommending possible organizational structures. The next day, this PSAC study, which came to be known as the Purcell Report after its chairman Edward Purcell of Harvard, was initiated and publicly announced. Eisenhower made known his strong preference for keeping civil space efforts within ARPA during a private meeting with top GOP congressional leadership also held on the fourth of February. Eisenhower wanted to keep his top-priority WS-117L program shielded and on track while avoiding the duplication he saw arising from creation of a civil space agency; Killian and Vice President Nixon immediately objected to Eisenhower's approach arguing, respectively, that "a truly scientific space aspect does exist" and that the U.S. position in world opinion would benefit "if non-military research in outer space were carried forward by an agency entirely separate from the military" (Divine 1993, 101). When Killian outlined the recommendations of the Purcell Committee in a memorandum to Eisenhower on March 5, the President now responded enthusiastically to the plan to create a civilian space agency and Killian supervised writing proposed legislation that was delivered to Congress on April 2 following a brief and minimal interagency coordination process.

Despite the strong congressional interest in rapidly creating a civilian space agency, it soon became clear that Congress had no intention of simply rubber-stamping the administration's proposal; both houses held extensive hearings on the proposal and soon drifted into positions that differed from one another and from the administration. The most significant debates were in three issue-areas: the relative priority of the nation's civil and military space efforts, the relationship between civilian and the military space organizations, and decision-making structures for creating overall national space policy. Substantially different bills were sent out of the House and Senate: the House bill emphasized the priority of civil space, favored NASA over DoD, and provided for a space advisory committee on the Atomic Energy Commission model; the Senate version indicated that DoD must remain independent of NASA on military space issues and called for the creation of a high-level space-policy decision-making body on the NSC model. The deadlock between the two bills was resolved after Eisenhower met with Senator Johnson on July 7 and agreed to accept Johnson's NSC-type committee if the President was made chairman of the committee. At a conference committee meeting on July 15 further compromises allowed a modified version of the House's Civilian-Military Liaison Committee between NASA and DoD, creation of the National Aeronautics and Space Council (NASC) at the White House, and the final wording of Section 2 indicating that NASA would exercise control over U.S. space activities "except that activities peculiar to or primarily associated with the development of weapons systems, military operations, or the defense of the U.S. [including the Research and Development necessary to make effective provision for the defense of the U.S.] shall be the responsibility of and shall be directed by the DOD" (Schoettle 1966, 260–261). Both Houses immediately passed the conference bill, Eisenhower signed the National Aeronautics and Space Act into law on July 29, and NASA was established on October 1, 1958.

Creation of NASA marked the formal beginning of two separate, but closely related and imprecisely delineated, American space programs. NASA's rapid creation and its broad powers reflected the desire for a new civilian space organization and primary importance of the civil space mission. At the same time, the process of creating NASA also highlighted the perceived role of space in national security and the tone of the National Aeronautics and Space Act expressed these latter considerations to greater degree than had the administration's original proposal. The testimony of OSD and Air Force witnesses was an important input to the political process by which these security considerations had been voiced more strongly than Killian and the Purcell Committee had originally intended. In bureaucratic terms, the creation of NASA at least solidified and perhaps even enhanced the military space positions of ARPA, OSD, and the Air Force; the great loser was the Army, which was left with few military or civil space missions.

Frictions on several specific space issues, including control over the man-in-space mission, emerged between ARPA and NACA even before the National Aeronautics and Space Act was signed. In August 1958, Quarles and Killian reached a compromise that directed NASA to design and build the capsules for manned spaceflight while ARPA would continue to concentrate on the boosters required for this mission. NASA had been created to become the nation's primary space-exploration organization and had inherited NACA's infrastructure; but, despite its charter, NASA lacked specific space expertise in many areas, especially in booster development. The von Braun team at ABMA, by contrast, was the nation's leading booster development group and had been tasked by ARPA to study and design a 1.5 million pound thrust booster known as Saturn C-1, but lacked a specific military rationale for building this huge booster. In October, Quarles and Keith Glennan, NASA's first administrator, tried to resolve this anomalous situation by transferring the Army-sponsored Jet Propulsion Laboratory (JPL) at the California Institute of Technology and the von Braun team at ABMA to NASA. ABMA Commander Medaris and Army Secretary Wilber Brucker vigorously fought against this proposal and the NASC had to work out a compromise allowing JPL to become part of NASA and the von Braun team to remain under ABMA but work on Saturn under contract to NASA.

The continuing struggle of ABMA to remain a major player in the national space program and to retain control over the von Braun team next came to a head in the summer and fall of 1959. In an attempt to find a better rationale to promote their space capabilities, in March the Army organized a study of military uses for the Saturn booster known as Project Horizon. The Project Horizon Report was completed in June and detailed a comprehensive plan to establish a 12-man lunar outpost by November 1966; construction of the lunar outpost was to be supported by 149 Saturn launchings and estimated to cost $6 billion. Project Horizon and the Army's other overly ambitious attempts to retain its space expertise did not achieve their desired effect as Kistiakowsky, Quarles, DDR&E Herbert York, and eventually Eisenhower all became convinced that ABMA should be put under NASA. Congress approved this realignment in January 1960 and on July 1, 1960, Eisenhower presided over the opening of NASA's Marshall Space Flight

Center in Huntsville, Alabama. With the loss of JPL and the von Braun team, the Army no longer had the expertise or stomach to pursue a major space program and cleared the way for Air Force dominance of military space within DoD. The demise of the Army space program along with the rise of NASA led to a fundamental change in the character of the U.S. military space program away from a broader national approach toward a more narrow focus on military space. Medaris and von Braun had proposed the most far-reaching plans for exploration and use of space of any service; Army traditions led it to view its role in space within a broad exploratory context—more in terms of the Lewis and Clark Expedition or the work of Topographical Engineers in surveying routes for a transcontinental railroad than solely within a narrow warfighting context.

The *Sputnik* challenge illustrated the international prestige aspects of the opening of the space age, but the superpowers were also very concerned with the security implications of this new medium and jockeyed for position in attempting to present their space programs to the international community in the best light. Eisenhower initiated a series of exchanges with Soviet Premier Nikolai Bulganin in a January 1958 letter proposing the superpowers agree to use space for "peaceful purposes" only. The Soviets offered a sweeping response including a United Nations (UN) agency to launch spacecraft and supervise space; after much procedural maneuvering, on November 24, 1958, the UN Ad Hoc Committee on the Peaceful Uses of Outer Space (COPUOS) was created. The limited charter of COPUOS, the decision of the superpowers to posture rather than seriously address many substantive space issues, and adoption of the American distinction between space for "peaceful" rather than explicitly "nonmilitary" purposes all served to guard U.S. military space programs and set the international law context for the early space age.

The United States adopted its first comprehensive space policy in a secret document, NSC 5814/1, "Statement of Preliminary U.S. Policy on Outer Space," on August 18, 1958. The report skirted the issue of where space begins by stating that "the upper limit of air space has not been defined" but clearly conflicted with the Air Force's emerging aerospace concept: "For the purposes of this policy statement, space is divided into two regions: 'air space' and 'outer space.'" It divided military space programs into three chronological periods: "Now Planned or in Immediate Prospect" containing military reconnaissance; "Feasible in the Near Future" with weather, communications, navigation, and electronic counter-measures satellites; and "Future Possibilities" containing manned maintenance and resupply of space vehicles, manned defensive vehicles, bombardment satellites, and manned lunar stations. Turning to manned spaceflight, the statement indicated that manned exploration would represent the true conquest of space and noted that unmanned missions could not substitute in terms of their psychological effect on the world public. Finally, the report also emphasized that *"Numerous legal problems* will be posed by the development of activities in space" but that *"rules will have to be evolved gradually"* and that space *"is not suitable for abstract a priori codification"* (U.S. Dept of State 1958–1960, 834–840, emphasis in original).

The space for peaceful purposes legal and policy environment allowed the United States to experiment with ground-based capabilities suitable for attacking or defending satellites. In August and September 1958 and again during the summer and fall of 1962, the United States conducted a series of high-altitude nuclear detonations that were designed to test and did confirm that high-energy electrons produced in such detonations would become trapped in the Earth's magnetic field. The pumped up radiation belts caused by the 1.4 megaton Starfish Prime detonation 248 miles above Johnson Island in July 1962 caused premature failure of all seven satellites then in low-Earth orbit (LEO) as well as significant communications disruptions in Hawaii, some 700 miles away. The most comprehensive and largest initial anti-satellite (ASAT) R&D effort was a proposed satellite interception and inspection system known as SAINT; studies were initiated in the late 1950s but the program was canceled in December 1962 before any development or testing. Under Project Bold Orion in October 1959, the United States conducted the world's first ASAT test by using a B-47 bomber to air-launch a missile that passed within four miles of the *Explorer* VI satellite. The United States also deployed two nuclear-tipped ASAT missiles in the Pacific during the 1960s. The Army's Program 505 was a modified Nike Zeus missile stationed at Kwajalein Atoll that became operational in August 1963 and was deactivated by 1967. The Air Force's Program 437 Thor ASAT was located on Johnson Island and operational from June 1964 until the early 1970s. Both of these initial ASAT systems suffered from a number of significant operational deficiencies including: an inability to attack many satellites in different types of orbits due to the range and azimuth limitations imposed by the missiles themselves and by having only two launch sites for these direct-assent systems; an inability to discriminate in attacking individual targets due to the nuclear warhead; and a limited number of ASAT missiles, inadequate tracking and targeting support, and a weak logistical infrastructure.

In September 1959, DDR&E York initiated evolutionary changes in the DoD space structure that gave the Air Force control over virtually all DoD booster development programs and space launches, then in March 1961 he helped make the Air Force responsible for most DoD space R&D. York's changes were designed both to consolidate military space activities and encourage greater Air Force responsibility in proposing space plans by forcing budgetary tradeoffs between space programs and all other Air Force programs. This maneuvering to consolidate the Air Force's space position by York caused consternation within the other services since it weakened ARPA and each of the services also wanted some part in the significant military man-in-space mission that now seemed imminent. Beginning in the summer of 1959, the Army and Navy, for the first time, seriously proposed the creation of a unified or joint-service space command with the responsibility for development and production of all space vehicles and boosters. The Chief of Naval Operations (CNO), Admiral Arleigh Burke, forwarded a unified space command proposal from the Joint Chiefs of Staff (JCS) to the Secretary but York and McElroy ruled against the Army and Navy. McElroy's September 18 response memorandum found that "establishment of a joint military organization with control over operational space systems does not appear desirable at this time" (U.S. Congress 1961, 10; see also Spires 2007, 76–78).

Air Force control over military space activities became even more secure shortly after the arrival of the Kennedy Administration. During the transition period President-elect Kennedy had asked Jerome Wiesner of MIT (who would become his Science Adviser) to head the Ad Hoc Committee on Space and tasked this group to study the structure for and direction of U.S. space efforts. Wiesner's report recommended a revitalization of the NASC, called for primary emphasis on space-science missions, warned against attempting to race the Soviets for manned space spectaculars, and strongly opposed DoD's overlapping space programs and duplication of NASA's work. The new Secretary of Defense, Robert McNamara, agreed with the tenor of the report's recommendations and tasked his office to begin the review of military space organizations that led to Defense Directive 5160.32, "Development of Space Systems," issued on March 6, 1961. This directive indicated that each service or defense agency would be allowed to conduct "preliminary research to develop new ways of using space technology to perform its assigned function" subject to guidelines established by the DDR&E and that all approved space projects then became the responsibility of the Air Force (Stares 1985, 60–61).[2] The Army and the Navy were allowed to retain their primary space programs, the Army Advent satellite communications system and the Navy Transit navigation satellite system, but all future space R&D programs were subject to the new procedure. McNamara viewed this directive as a way to exert more direct and tighter control over DoD space efforts and believed that centralizing DoD space R&D within the Air Force was the easiest way to gain greater control.

Encouraged by the space rhetoric of the Kennedy campaign and the arrival of Vice President Johnson as chair of the NASC, the Air Force attempted to build its case for an expanded military space presence. In October 1960, General Schriever had asked Trevor Gardner to chair the Air Force Space Study Committee and examine future military options in space. The top secret Gardner Report was completed in March 1961 and provided a ringing endorsement of the high-ground and space-control schools already prevalent within the service. The report basically ignored NASA and called for the new Air Force Systems Command (AFSC) to spearhead an accelerated and very ambitious program including manned spaceflight, space weapons, reconnaissance systems, large boosters, space stations, and even a lunar landing by 1967–1970.

Rather than this preferred Air Force option, the Kennedy Administration advanced three other major initiatives in space policy: the top-priority prestige-based Moon landing race with the Soviets, the space-related arms-control process that would culminate in the 1967 Outer Space Treaty (OST), and secret efforts to protect and further legitimize the emerging spy-satellite regime. With Yuri Gagarin's successful orbital flight on April 12, 1961, the Soviet Union scored another major space first that was again a propaganda tool for touting the superiority of their system. Kennedy felt compelled to show America's strength by beating the Soviets in some race to demonstrate U.S. science and technology prowess, tasking Johnson to determine which spectacular offered America the best chances for victory and prestige. Johnson quickly centered on a Moon race as the best response to the Soviets and his recommendation led directly to Kennedy's May 25 Moon-landing challenge. Congress enthusiastically embraced the challenge, which quickly became Project Apollo, the single largest and most important U.S. space program to date. In a very real sense, the final U.S. response to *Sputnik* was not complete until Neil Armstrong and Buzz Aldrin walked upon the Sea of Tranquility on July 20, 1969. As America's predominant space effort during the 1960s and early 1970s, the Moon race completely overshadowed all other U.S. space activities such as the Air Force's continuing attempts to justify and develop a manned military space mission. As NASA's budget grew from $964 million in 1961 to $5.1 billion by 1964 while the DoD space budget went from $814 million to $1.6 billion for the same period, fears that the DoD would somehow dominate or subvert NASA were completely erased.

Another major space-policy focus during the Kennedy Administration was a quest for arms control in space. The administration took a "two-track" approach to ASAT development and arms-control efforts by deploying a minimum number of ASATs to mitigate against a Soviet orbital nuclear-weapon threat while simultaneously pursuing arms-control efforts to ban such weapons in space and thereby removing a major incentive for deploying ASATs. Early in the Kennedy Administration, however, efforts to achieve space arms control were severely hampered by a lack of interagency coordination. Accordingly, on May 26, 1962, Kennedy issued National Security Action Memorandum (NSAM) 156—an implicit critique of the NASC and a request for the Department of State to create a high-level coordinating body (known as the NASM 156 Committee) to address this problem. The primary responsibility of the NASM 156 Committee was to develop policies designed to protect and legitimize U.S. spysats, but this group was also chiefly responsible for creating the U.S. initiatives aimed at banning nuclear weapons from outer space. During the summer and fall of 1962, the NSAM 156 Committee was the scene of intense interagency disputes with the State Department and the Arms Control and Disarmament Agency (ACDA) most supportive of banning nuclear weapons from space and the JCS most strongly opposed, arguing that a ban could not be verified and would preclude military options. Interagency bargaining set in motion the informal and formal initiatives that eventually led to the international declaratory ban on placing nuclear weapons or weapons of mass destruction in outer space expressed in UN General Assembly (UNGA) Resolution 1884 on October 17, 1963. It is noteworthy that America's approach to space arms control marked the first time the U.S. government was willing to initiate arms-control discussions with the Soviet Union that did not make inspection or verification a prerequisite. This willingness to obtain an unsecured agreement with the Soviets was an illustration of the administration's general de-emphasis on military space programs in favor of peaceful and civil uses of space as well as an expression of their judgments that nuclear weapons in space: lacked military utility, were not required by U.S. military space doctrine, and were better dealt with through a declaratory ban than via ASAT weapons.

The final major space-policy initiatives of the Kennedy Administration continued efforts to hide and legitimize spysat operations. Beginning in 1961, a security clampdown was slowly implemented, first on spy-satellite programs and then on all military space efforts. The final step in this process was the classified DoD "Blackout" Directive issued on March 23, 1962, that prohibited advance announcement and press coverage of all military space launches, forbade use of the names of military space programs such as Midas and Samos, and replaced program names with program numbers. To legitimize spysat operations further in international law, the administration used the UNCOPUOS and introduced a proposal for all states to provide data on their space launches to the UN that became the basis for UNGA Resolution 1721 adopted on December 20, 1961. Despite State's formal coordination of this proposal, it had completely bypassed the new and highly secret NRO, an oversight that highlighted the need for better interagency coordination on space issues and contributed to NSAM 156. As a Soviet diplomatic offensive against U.S. spysats reached its crescendo at the UN and elsewhere during 1962, the NSAM 156 Committee worked to tighten and strengthen the diplomatic and public rationale for spysats, primarily by emphasizing the peaceful purposes tenets of Eisenhower's space policy. In a key speech on December 3, 1962, at the UN, Ambassador Albert Gore asserted that:

> It is the view of the United States that Outer Space should be used for peaceful—that is non-aggressive and beneficial—purposes. The question of military activities in space cannot be divorced from the question of military activities on earth.
>
> There is, in any event, no workable dividing line between military and non-military uses of space. One of the consequences of these factors is that any nation may use space satellites for such purposes such as observation and information gathering. Observation from space

is consistent with international law, just as observation from the high seas. (Stares 1985, 70–71)

The Soviets began dropping their objections as their own spysats began to come on line beginning in the fall of 1962.

For much of the early space age the military desired to build a large manned military presence in space and often assumed that such a presence in space would be established soon. Based upon historical analogies, the military believed that man was an essential part of any program designed to exert control over space or to exploit the high-ground potential of space. The Air Force was at the vanguard of this outlook and developed the first comprehensive U.S. plans for manned space programs. The Air Force pushed hard to obtain approval for its manned space program prior to the establishment of NASA, releasing an ambitious plan in April 1958 that called for a first orbital flight by April 1960 and a first lunar landing by December 1965. Killian and NACA Director Hugh Dryden strongly opposed the Air Force's plan and urged the President to make NASA primarily responsible for the manned mission; their argument matched with Eisenhower's policy about space for peaceful purposes and his desire to avoid costly space races such as the Air Force proposed. Accordingly, Eisenhower formally gave NASA primary authority over U.S. manned spaceflight efforts in August 1958 and by November this effort had evolved into Project Mercury.

The Air Force continued to campaign for a large military space mission by shifting focus away from the race-driven non-maneuverable manned capsule approach it had earlier proposed (an approach derided as "Spam in a can") toward the more militarily useful Dynamic Soaring or Dynosoar piloted approach. In the early 1960s, McNamara's systems analysts showed that a modified Gemini capsule might perform military functions better and more cheaply than the Dynosoar and this prompted McNamara to attempt to gain a large role or control over this program, a move that NASA Administrator James Webb successfully parried by citing the impact of such a restructuring on the nation's highest-priority Apollo Program. Instead, in January 1963, Webb and McNamara signed an agreement to allow DoD experiments on some Gemini missions in a program known as Blue Gemini. This approach placed additional pressures on the Dynosoar (now known as X-20) and contributed to perceptions that it was very expensive and held limited military potential. As the Kennedy administration progressed, the Air Force fought an increasingly difficult losing battle to establish a manned military space presence due to OSD's building block approach and the constraints of the security blackout described above. The end of these early Air Force manned military spaceflight dreams came on December 10, 1963, when Secretary McNamara canceled the X-20 and at the same time continued the building block approach to determine the military utility of human spaceflight by assigning primary responsibility for developing the Manned Orbiting Laboratory (MOL) to the Air Force.

The Moon Race and Sanctuary Supreme, 1964–1980

The 1967 OST is the most important space-related arms-control agreement and is the foundation for the current space governance regime. By 1963, the Kennedy Administration had concluded that the United States could achieve significant space-related national security objectives via arms control and the NSAM 156 Committee had developed the U.S. negotiating positions that laid the groundwork for the OST including UNGA Resolutions 1884 and 1962 in October and December 1963: an international declaratory ban on placing nuclear weapons or weapons of mass destruction in outer space and a declaration that space was to be free for exploration by all, would not be subject to appropriation by national sovereignty, and that launching states would maintain jurisdiction over spacecraft and be liable for damage they might cause. On April 5, 1966, National Security Adviser Walt Rostow wrote a memorandum to President Johnson recommending the United States rapidly propose an international treaty at the UN to codify the principles in UNGA 1962 before the Soviets tabled their own draft treaty on this issue, and President Johnson publicly outlined the basic provisions of the U.S. draft treaty on May 7. Following negotiations and adoption of some provisions from the Soviet draft treaty, the UNGA endorsed the agreement on December 17 and on January 27, 1967, the Treaty on Principles Governing the Activities of States in the Exploration and Use of Outer Space, Including the Moon and Other Celestial Bodies was open for signature. Sixty-two states initially signed the OST and the agreement went to the U.S. Senate for advice and consent to ratification on February 7.

Many provisions of the OST echo UNGA Resolutions 1884 and 1962. The treaty purports to contribute to broad international cooperation in scientific and legal aspects of exploration and use of space for peaceful purposes, reaffirms the principle of freedom of use of space, makes activities in space subject to international law including the UN Charter, and stipulates that space is the province of all mankind and its use and exploration shall be carried out for the benefit and in the interest of all states. In addition, the treaty indicates that space, including the Moon and other celestial bodies, is not subject to national appropriation by claim of sovereignty, occupation, or any other means; signatories are not to place in orbit around Earth any objects carrying nuclear weapons or any other kinds of weapons of mass destruction (WMD), install such weapons on celestial bodies, or station such weapons in space in any other manner; and are forbidden to establish military bases, installations, or fortifications, test any type of weapon, or conduct military maneuvers on celestial bodies. Several issues were clarified in the detailed public discussions during the Senate hearings prior to the Senate's 88–0 vote in support of ratification on April 25, 1967, including, assurances U.S. national technical means (NTM) of verification could detect larger-scale deployments of nuclear weapons or WMD in space before they became militarily significant, U.S. preference to rely on its NTM for verification rather than attempting to create an international on-site inspection regime for objects in space, and Secretary of State Dean Rusk's assertion that the "treaty does not inhibit, of course, the development of an anti-satellite capability in the event that should become necessary" (U.S. Congress 1967, 26). Later in 1967, Secretary McNamara revealed that the Soviets had been testing a new type of ballistic missile delivery system known as a fractional orbital bombardment system (FOBS) that would bypass much of the U.S. strategic warning network but downplayed the significance of this system, stating it did not pose a major new strategic threat or violate the OST since the payloads were not in sustained orbit.

Announced on December 10, 1963, the MOL quickly took the place of the X-20 and became the Air Force's most important space program as the cornerstone of its efforts to build a significant military spaceflight presence. By mid-1965, the proposed design had turned the MOL into a formidable reconnaissance outpost with a large 90-inch reconnaissance telescope and huge signals intelligence antennas to be assembled on orbit alongside the station. The MOL design called for a configuration consisting of a Gemini

B capsule attached to the 41-foot-long laboratory that was to be launched into polar orbit from Vandenberg AFB atop a Titan III-C booster. The program was originally scheduled to include five manned MOL flights beginning in 1968 at a cost of $1.5 billion. The overall objectives of the MOL program were to: learn more about what man is able to do in space and how that ability can be used for military purposes and develop and experiment with technology and equipment that will help advance manned and unmanned spaceflight. These objectives dovetailed with the Air Force's belief that human spaceflight was key to its future in space and that the most important military space missions would become feasible only by having man's discriminatory intelligence on orbit. In February 1969, incoming Secretary of Defense Melvin Laird endorsed a comprehensive review of the MOL program that supported the program and found it could meet all its objectives with only six launches (two unmanned and four manned), but budgetary pressures and the ongoing war in Vietnam prompted the Nixon Administration to cancel the program on June 10, 1969. A total of $1.4 billion was spent on the MOL program, making it one of the most expensive military programs ever prematurely terminated as of that date.

A final very significant factor in the demise of the MOL program was the growing belief that unmanned spysats could perform the primary mission of the MOL as well or better at lower cost. NRO had been leery of the idea of a manned reconnaissance system from the outset, reasoning that a manned system might present more of a provocation to the Soviets, that the contributions of manned operators in space would not be significant when balanced against the costs and requirements of life support systems, and that any accident involving MOL astronauts might set back all space-based intelligence gathering unacceptably. As NRO began developing the United States' fourth generation photoreconnaissance satellites known as the KH-9 or "Big Bird," it also advanced the argument that the projected capabilities of the KH-9 would make MOL unnecessary.

Signed on May 26, 1972, SALT I consisting of the Treaty on the Limitation of Anti-Ballistic Missile Systems (ABMT) and the Interim Agreement on the Limitation of Strategic Offensive Arms heralded the arrival of the era of detente between the superpowers and was an attempt to codify mutual assured destruction (MAD) as the putative basis for strategic stability. The MAD paradigm had been largely conceptualized and developed by Secretary

McNamara in the mid-1960s; it posits that strategic forces beyond those required for assured destruction are not politically or militarily useful because a plateau of strategic stability can be achieved when each superpower possesses invulnerable second-strike strategic forces capable of delivering assured destruction and its urban-industrial targets are left undefended. There were important space-policy-related components to SALT I: the central role of spysats in enabling these negotiations, in verifying the agreements, and in motivating the legitimization of NTM, as well as prohibitions on space-based ABM systems in the ABMT. Under Article XII of the ABMT the superpowers were to use NTM to provide assurance of compliance with treaty provisions, and they pledged not to interfere with NTM performing this function or use deliberate concealment measures that would impede verification. More subtly, SALT established a direct interrelationship between NTM capabilities and the units of limitation in arms-control agreements, since these limits could only be as precise as could be "seen" by NTM. Examples of this relationship in practice include use of Vela Hotel satellite monitoring (the first NTM) to add space to the areas where nuclear testing was prohibited under the 1963 Limited Test Ban Treaty and the improvements in capabilities implied by the differences in the units of limitation between SALT I and SALT II: in 1972, NTM was asked to count very large immobile objects such as missile silos and Large Phased-Array Radars; by 1979, NTM was expected to be able to distinguish between types of ICBMs and to count numbers of warheads. In addition, the protections for NTM first adopted in Article XII of the ABMT and repeated in a number of subsequent agreements as well as the practice of the signatories undoubtedly establish some degree of protection for spysats in customary international law. Under a positivist interpretation of this language and the given opacity of many space operations, however, it is questionable just how much protection or legitimization it provides for NTM or satellites more generally since the language clearly stops well short of being a blanket ASAT weapons ban or even a clear approval of all spying from space.

Finally, the ABMT raises major issues about its unclear prohibitions on space-based ABM systems and the potential for space-based defense in general. Initially, specific prohibitions on space-based ABM systems found in Article V seemed clear: signatories were not to develop, test, or deploy ABM systems or components which are sea-based, air-based, space-based, or mobile land-based. Debate ensued during the Reagan Administration, however, when it asserted that under the "broad" or "legally-correct interpretation," based on the Treaty's Agreed Statement D, the parties were free to develop and test any type (space-based, mobile land-based, etc.) of ABM system (at locations other than the agreed test ranges) so long as these new ABM systems were based on other physical principles. Citing the changed strategic environment following the end of the Cold War, in June 2002 the United States exercised its option to withdraw from the ABMT.

During the 1960s and 1970s, the United States first deployed comprehensive military space capabilities including battlefield and ocean surveillance, communication, navigation, meteorological, mapping, charting, and geodesy satellites that have been updated and are still operational today. These new and improved satellite systems for early warning and force enhancement allowed space data flows to become more routine and reliable, thereby enhancing the effectiveness of terrestrial forces tremendously. Vastly improved force enhancement capabilities created a nascent reconnaissance-strike complex, sparking a revolution in space and terrestrial military operations that was little understood at the time and was not the product of proactive, top-down guidance but rather, relatively unfocused, bottom-up, and incremental technical improvements.

The United States also deployed its third, fourth, and fifth generation U.S. photoreconnaissance satellites as well as several generations of sophisticated signals intelligence (SIGINT) satellites during this time. At the beginning of this period, three descendants of the Corona and Samos programs first came on line, the area surveillance KH-4, and the multi-spectral, close-look capabilities of the KH-7 and KH-8. Like the Corona system before them, all these systems returned film imagery in reentry capsules and were unable to provide timely information on fast-breaking events such as the 1968 Warsaw Pact invasion of Czechoslovakia or the Arab-Israeli wars in 1967 or 1973. Continuing debates over the allocation of spysats to specific missions and over differences in data interpretation led to the creation of the Committee on Imagery Requirements and Exploitation (Comirex) in 1967. Comirex was responsible for prioritizing and scheduling available intelligence assets against desired targets and also assigned primary responsibility for data interpretation on specific targets to individual agencies. The revolutionary fifth generation electro-optical system known as the KH-11 first launched in December 1976 solved the timeliness problem with a near-real time direct data downlink.

Cumulatively, the technological wonders described above and the policies for their use had a profound impact on U.S. military space doctrine and perceptions toward using space during this period. By the time he became President, Lyndon Johnson had moved far away from his initial position on space as the new high ground. Consider his often cited off-the-record remarks to a group of Tennessee educators in March 1967:

> I wouldn't want to be quoted on this but we've spent 35 or 40 billion dollars on the space program. And if nothing else had come out of it except the knowledge we've gained from space photography, it would be worth ten times what the whole program has cost. Because tonight we know how many missiles the enemy has and, it turned out, our guesses were way off. We were doing things we didn't need to do. We were building things we didn't need to build. We were harboring fears we didn't need to harbor. (Quoted in Burrows 1986, vii)

President Johnson's remarks refer directly back to the missile-gap episode and reflected his faith in and enthusiasm for space-based reconnaissance. These sentiments were instrumental in moving Johnson to propose at the Glassboro Summit three months later the process that initiated superpower arms-control negotiations and led to the SALT I agreements.

The question of next appropriate goals following its triumph in the Moon race was the overriding issue for U.S. space policy in the late 1960s and early 1970s. Shortly after entering office, President Richard Nixon established a Space Task Group (STG) to complete a comprehensive review of the future plans of the U.S. space program comprised of Vice President Spiro Agnew, Acting NASA Administrator Thomas Paine, Secretary of Defense Laird, and Science Adviser Lee DuBridge. On September 15, the STG presented Nixon with three options for post-Apollo U.S. civil space plans. Option one called for a manned mission to Mars by 1985 supported by a 50-person space station in Earth orbit, a smaller space station in lunar orbit, a lunar base, a space shuttle to service the Earth space station, and a space tug to service the lunar stations. Option two consisted of all of the above except for the lunar projects and delayed the Mars landing until 1986. Option three included only the space station and the space shuttle, deferring the decision on a Mars mission but keeping it as a goal to be realized before the end of the century. The report estimated that annual costs for option one would be $10 billion, option two $8 billion, and option three $5 billion. Considering that NASA's budget had peaked at the height of the Moon race in 1965 at a little more than $5 billion and that political support for space spectaculars was rapidly eroding, the STG recommendations seemed fiscally irresponsible and politically naive. On the basis of the STG report and the recommendations from other major space studies during this period, Nixon formalized U.S. post-Apollo space-policy goals in March 1970 by endorsing development of a shuttle and leaving a space station or Mars mission contingent upon successful completion of a shuttle program.

Interactions between NASA and DoD were important in the design and operations of the Shuttle program. NASA's decision to pursue a large shuttle necessitated a design able to accommodate the most important potential users and satisfy the military in particular. The Air Force set a number of challenging performance criteria including a requirement to lift a 65,000 pound payload and moved NASA toward a lifting body design known as the Thrust-Assisted Orbiter Shuttle (TAOS) that was formally approved as the Space Transportation System (STS) by Nixon on January 5, 1972. When the STS ran into great political and budgetary problems during the Carter Administration, DoD stepped in to help save the program—largely due to the shuttle's projected capability to launch huge spy satellites. Thus, the rationale behind the STS became increasingly militarized and related to spysats. Shuttle capabilities also allowed the military again to entertain plans to develop a manned military presence in space.

The initial spaceflight of *Columbia* on April 12, 1981, marked a bittersweet milestone because it was the world's first reusable spacecraft and signified the return of American human spaceflight, but the STS was also two years behind schedule and cost $2 billion more to develop than originally projected. During the mid-1980s, NRO was eventually allowed to become the only government entity to build a backup launcher, the Complementary ELV (CELV), for its most important payloads. Following the *Challenger* disaster on January 28, 1986, NASA, DoD, and the newly formed Office of Commercial Space Transportation within the Department of Transportation produced a new U.S. Space Launch Strategy and the Space Launch Recovery Plan. The plan specified that the United States would rely on a balanced mix of STS and expendable launch vehicles (ELVs) with critical payloads designed for launch by either system.

Following the limited capability provided by deployment of Program 505 and 437 ASAT systems in the 1960s, neither DoD nor civilian leadership showed interest in maintaining these systems or pursuing new ASAT capabilities; by the early 1970s these systems were no longer operational. Meanwhile, the Soviets developed a nonnuclear co-orbital ASAT that was initially tested at least seven times between 1968 and 1971. This testing as well as recognition of the growing significance of military space systems and their vulnerability prompted the United States to look again at the need for an ASAT system. In February 1976, the Soviets resumed testing their co-orbital ASAT system using a different seeker, beginning a 13-launch test series lasting until June 1982. Following a number of studies for the NSC staff and confirmation from DoD that efforts to remedy U.S. satellite vulnerability had not proceeded very far, in the fall of 1976 President Gerald Ford issued National Security Decision Memorandum (NSDM)-333 directing DoD to increase its efforts and funding to reduce its satellite vulnerability problems. The NSC studies concluded that a U.S. ASAT would not enhance the survivability of U.S. satellites by deterring use of the Soviet ASAT because the United States was more dependent upon space than the Soviets. The reports also concluded, however, that a U.S. ASAT could be used to counter the threat to U.S. forces posed by Soviet space-based targeting systems and that development of a U.S. system could serve as a "bargaining chip" in possible U.S.-U.S.S.R. ASAT arms-control negotiations. On January 18, 1977, President Ford issued NSDM-345 authorizing a new nonnuclear ASAT, leading to eventual development of the air-launched miniature homing vehicle (MHV).

Jimmy Carter was the first President since Eisenhower to promulgate a National Space Policy. Presidential Directive (PD)-37, "National Space Policy," was a secret document signed by Carter on May 11, 1978. In June, a White House fact sheet explained the main provisions, including work to resolve potential conflicts among space program sectors; use of the STS for all authorized customers, domestic, foreign, commercial and governmental; and an aggressive, long-term program to provide greater survivability for military space systems. National Security Adviser Zbigniew Brzezinski signed PD-42, "Civil and Further National Space Policy," on October 10, 1978, and a fact sheet was released the next day. The fact sheet indicated three components for U.S. civil space

policy: (1) evolutionary space activities unique to or more efficiently accomplished in space; (2) a balanced strategy of applications, science, and technology development; and (3) emphasis that
it was neither feasible nor necessary to initiate a challenging space
program comparable to Apollo at this time. Overall, Carter's space
policy reflected both the traditional uneasiness of the scientific
community with space spectaculars and recognized the growing
importance of space across all sectors.

President Carter attempted a two-track approach by entering into ASAT negotiations with the Soviets while simultaneously
continuing development of the MHV ASAT authorized by Ford.
U.S. and Soviet negotiators met for three rounds of ASAT talks:
June 8–16, 1978, in Helsinki; January 23–February 16, 1979, in Bern;
and April 23–June 17, 1979, in Vienna. Like all serious international
negotiations, these talks were conducted in secret and the negotia-
tion record has not yet been declassified but it is clear the two sides
never came close to concluding an agreement. Controversies that
have publicly emerged include debates over: an ASAT ban versus
limitations or "rules of the road"; the degree of protection afforded
to third-party satellites; long versus short testing moratoria; and
how to deal with systems having residual ASAT capabilities—for
example, the Soviets insisted that the U.S. STS then under development be included as an ASAT system. Both sides expected that
the negotiations would continue but they never restarted due to
several factors including the breakdown of relations following
the Soviet invasion of Afghanistan and the arrival of the Reagan
Administration with its initial lesser enthusiasm for arms control.
Paul Stares argued that the U.S. two-track approach to ASAT arms
control legitimized and perpetuated the MHV ASAT system—a
system he believes had value only as a bargaining chip. The ASAT
negotiations also highlight what Ashton Carter refers to as "the
basic paradox of ASAT arms control": an inverse relationship between ASATs and incentives to place very threatening military
systems in space. Clearly, space weapons cannot be divorced from
terrestrial security considerations, the natural offense-defense dialectic, and the trade-offs inherent in all strategic thinking.

Space Comes of Age, 1981–Present

The Reagan Administration devoted considerable attention to space
issues and generated more official space-policy statements than

any other administration during the Cold War. Reagan's space-policy statements covered a wide range of topics but generally emphasized military space potential and space commercialization efforts to a greater degree than any previous administration. The Reagan Administration's first major space policy was contained in National Security Decision Directive (NSDD)-42 and publicly announced on July 4, 1982. NSDD-42 was crafted in late 1981 and the first half of 1982 by the Senior Interagency Group or SIG (Space) chaired by Science Adviser George Keyworth. Much of Reagan's first comprehensive space policy echoed Carter's space policies and also reflected the earliest principles of U.S. space policy established under Eisenhower. For example, NSDD-42 continued the U.S. emphasis on the "use of outer space by all nations for peaceful purposes and for the benefit of all mankind," rejected "any limitations on the fundamental right to acquire data from space," and reemphasized the U.S. positions that space systems are considered national assets and that "purposeful interference" with space systems "shall be viewed as an infringement upon sovereign rights." One significant change in emphasis between the Carter and Reagan policies concerned the perceived utility of space-related arms control: the Reagan Administration was far less enthusiastic and specifically indicated it would "oppose arms control concepts or legal regimes that seek general prohibitions on the military or intelligence use of space" (National Security Decision Directive Number 42, 1982, 2–3).[3] Due to major developments including the Strategic Defense Initiative (SDI) and Challenger disaster, the Reagan Administration issued a revised comprehensive directive on overall U.S. national space policy on January 5, 1988, along with White House fact sheets released on February 11, 1988. The new policy again exhibited great continuity with earlier space policies but did include some significant changes: maintenance of U.S. space leadership was no longer a basic goal, more emphasis was placed on promoting commercial space activities, and expanding human presence and activity beyond Earth orbit was added as a basic goal. The directive also reflected the four-part military space typology first used in DoD space-policy statements earlier in the 1980s: space support, force enhancement, space control, and force application. Under space support, DoD was directed to maintain launch capability on both coasts and to enhance the robustness of its satellite control capability. For force enhancement, DoD was to develop space systems and plans to support operational forces at all levels of conflict. In the space-control area, DoD was directed to develop an integrated combination of ASAT, survivability, and surveillance capabilities, including deployment of a robust and comprehensive ASAT capability at the earliest possible date. Finally, under force application, consistent with treaty obligations, DoD would conduct research, development, and planning to be prepared to acquire and deploy space weapons systems for strategic defense should national security conditions dictate. During the last year of the Cold War, the newly created National Space Council chaired by Vice President Dan Quayle developed the space-policy directive President George Bush signed on November 2, 1989. The unclassified "National Space Policy" released on November 16 was virtually identical to the February 11, 1988, fact sheet on the last Reagan space policy.

The most important space-related Air Force organizational development of the Cold War era was the creation of Air Force Space Command (AFSPC) on September 1, 1982. AFSPC was the first completely new major command formed by the Air Force in 32 years. The new command was designed to consolidate, centralize, and focus many of the Air Force's space efforts. AFSPC was also the product of many factors and was the result of significant internal bureaucratic struggles within the Air Force. Pressures to create a separate space command within the Air Force came both from the top-down and from the bottom-up. However, unlike almost all other military space-related issues, the decision to create a separate major command for space was mostly an internal Air Force bureaucratic matter. A unified command structure consisting of all services, United States Space Command (USSPACECOM), was established above AFSPC on September 23, 1985. On October 1, 2002, USSPACECOM was merged into United States Strategic Command (USSTRATCOM). Although this was originally described as a joining of equals, in practice this major organizational shift quickly amounted to the absorption of USSPACECOM into USSTRATCOM and left very few vestiges of the original USSPACECOM. Instead of space being the sole focus of one of just nine unified commands, under the new structure space now competes for attention among a very wide array of disparate mission areas that include deterring attacks on U.S. vital interests, ensuring freedom of action in space and cyberspace, delivering integrated kinetic and nonkinetic effects to include nuclear and information operations in support of U.S. Joint Force Commander operations, synchronizing missile defense plans and operations, and combating weapons of mass destruction. And because unified commands are the warfighters who operate systems and set capability requirements, this change has resulted in less focus on current space operations and future space capability needs.

A range of factors including difficulties in modernizing U.S. strategic nuclear forces, particularly in finding a suitable basing mode for the new MX "Peacekeeper" ICBM; growing strength of the nuclear freeze movement worldwide; and continuing Soviet strategic modernization combined to create strong incentives for the United States to reinvestigate the potential of strategic defenses, culminating in President Reagan's March 23, 1983, "Star Wars" speech that began the Strategic Defense Initiative (SDI). Reagan had been predisposed toward strategic defenses throughout his political career and moved strongly in this direction after receiving unanimous support for investigating new strategic defense possibilities at a White House meeting with Secretary of Defense Caspar Weinberger, National Security Adviser "Bud" McFarlane, and the JCS on February 11, 1983. By initiating a policy-push, top-priority, long-term research program to investigate the feasibility of strategic defenses, Reagan fundamentally altered the strategic landscape worldwide and in the words of SDI Historian Donald Baucom just may have given "the United States a second wind in the critical home stretch of the Cold War" (Baucom 1992, 198).

Reagan signed NSDD-85, "Eliminating the Threat from Ballistic Missiles," that formally began the initiative he had publicly announced two days earlier and soon thereafter, in National Security Study Directive (NSSD) 6–83, ordered two major studies on SDI be completed by October: The "Defensive Technologies Study," or Fletcher study after its chairman, former (and future) NASA Administrator James Fletcher; and the "Future Security Strategy Study" undertaken by two groups, an interagency team led by Franklin Miller and a team of outside experts chaired by Fred Hoffman. These reports found that many emerging technologies held significant missile-defense potential and recommended building defenses in multiple layers including space-based layers; they provided strong support for the authorization in NSDD-119 on January 6, 1984, to begin SDI and establish the SDI Office (SDIO) reporting to the Secretary of Defense later that month. On April 15, Lieutenant General James Abrahamson moved from his position as Associate Administrator for Manned Spaceflight at NASA to become the first director of SDIO. His new office first demonstrated the potential of emerging defense technologies on June 10 when a kinetic energy interceptor known as the Homing Overlay Experiment launched from Meck Island in the Pacific Test Range success-fully intercepted a test reentry vehicle launched atop a Minuteman ICBM from Vandenberg AFB, a test widely described as hitting a bullet with a bullet.

In an address on February 20, 1985, Special Arms Control Ambassador Paul Nitze officially codified two criteria by which SDI developments should be judged: first, that any defense systems be highly survivable, and second, that defense systems "be cost effective at the margin—that is, the incremental cost of adding additional defensive capability must be low enough so that the other side has little incentive to add additional offensive capability to overcome the defense" (Nitze 1985, 2). A series of SDIO experiments conducted on September 5, 1986, known as the Delta 180 test confirmed the ability of space-based infrared sensors and kinetic weapons to perform simulated boost-phase intercepts. During 1987, the JCS established classified minimum performance requirements for a Phase I strategic defense deployment and Secretary Weinberger announced on September 18 that the Defense Acquisition Board had formally moved six parts of the Phase I SDI program past the demonstration and validation milestone of the defense acquisition process. In a final major development at the end of the Cold War, Lowell Wood of the Lawrence Livermore National Laboratory proposed an inexpensive, autonomous, and small space-based kinetic interceptor system known as Brilliant Pebbles.

With the renewed interest in strategic defense during this period, the DSP system was seen as incapable of dealing with the launch detection and booster tracking needs of a BMD system. The Phase One SDI architecture called for replacing the DSP system with the 12-satellite boost surveillance and tracking system (BSTS) by the mid-1990s. Despite continuous improvements, the DSP system does not approach the type of comprehensive and highly accurate boost-phase coverage BSTS was designed to provide. Differences in capabilities between the DSP and proposed BSTS systems provide an excellent example of the impact of doctrinally derived mission requirements on space hardware. The DSP system was originally designed during the heyday of detente and MAD; despite being upgraded, it still performs best as a strategic "bellringer" system and has limited capabilities to support tactical and operational BMD applications such as closely determining launch sites or tracking boosters. Despite the very significant contributions of the DSP system during the Gulf War overall, tactical level weaknesses with the DSP were evident during the campaigns against Scud missile launchers. During the fall of 1990, the United States moved the two newest and most capable DSP satellites into geostationary Earth orbit (GEO) overhead the Gulf region to provide stereo imaging of missile launches from this area. This stereo imaging was critical in predicting impact points and providing at least 90 seconds of warning prior to Scud impact. This data also provided crucial cueing data to support Patriot interception attempts. The DSP system was less helpful in providing data to precisely locate the Scud launch sites for bombing by coalition forces. Reportedly, the DSP satellites could only localize Scud launch sites within approximately 2.2 nautical miles which, apparently, was not accurate enough to allow many of these mobile launchers to be destroyed by airstrikes during the course of the war.

Partially due to BMD requirements that might require a heavy lift capability, in May 1987 DoD initiated a joint program with NASA to develop a new ELV program known as the Advanced Launch System (ALS). The goals of the ALS program included the development of a flexible and reliable family of modular launch vehicles which could easily be configured for specific needs. The ALS was to use improved technology to lower the cost per pound to LEO initially to $1,500 and then to $300 by the late 1990s. Despite the apparent need for new booster technology and the possible need for heavy lift capability for SDI components and assembly of space station *Freedom*, the ALS program floundered and was no more than a study effort by the end of the Cold War. A second major joint DoD-NASA space launch technology development effort was emphasized by President Reagan in his 1986 State of the Union Address and was known as the X-30 or the National Aerospace Plane (NASP). The goal of the NASP program was to build an experimental manned single-stage-to-orbit vehicle which would take off like an airplane, fly into space, and then return to land like an airplane. The NASP program proved to be extremely challenging technologically and fiscally; the goal of building and testing a full-size, manned X-30 vehicle was abandoned in late 1992. A December 1992 Government Accountability Office (GAO) report estimated the total costs for a complete X-30 program would have been $17 billion versus the $3.1 billion original estimate in the late 1980s.

The Reagan Administration was the first systematically to use the products from spysats (and other sources) publicly in emphasizing the extent of the continuing Soviet military expansion as a part of its broader domestic and international efforts to build support for the U.S. defense buildup. From 1981 to 1989, DoD published and widely distributed an annual glossy report entitled *Soviet Military Power*. Remarkably detailed illustrations of almost all of the principal Soviet military systems were a major feature of these reports. As Jeffrey Richelson and others have pointed out, U.S. imagery satellites were undoubtedly the primary source for creating most of these illustrations. Presenting this spysat-derived data in a public report marked an important break with previous practice. In addition, there was slow but continuing movement toward developing programs to exploit these national spysat capabilities more effectively for military applications at the tactical level. Planning and training for certain highest-priority tactical military operations such as the 1970 raid on the Son Tay prisoner of war camp in North Vietnam and the 1980 Iranian hostage rescue attempt which ended at Desert One had relied heavily upon intelligence data collected from space-based assets; by the end of the Cold War the first sustained efforts to move beyond ad hoc efforts were established. The most prominent work toward this end is known as the Tactical Exploitation of National Capabilities or TENCAP, a program Congress directed the services to create in 1977.

The Air Force's MHV ASAT program was initiated in September 1977 and first flight tested in January 1984. The Reagan Administration faced rancorous and continuing high-level political debates over the U.S. strategic need for an ASAT, the prospects for ASAT arms control, and specific testing restrictions for the MHV ASAT. On July 18, 1983, Senator Paul Tsongas's amendment to the Fiscal Year (FY) 1984 DoD Authorization Act withheld funds for testing the MHV ASAT system unless the President certified both that the United States was negotiating with the Soviets in good faith about ASATs and that testing was needed for U.S. national security. The House Appropriations Committee went even further and withheld $19.4 million in FY 1984 advanced procurement funds from the MHV; following administration pressure, the conference committee restored funding with the proviso the administration provide Congress a report on U.S. ASAT policy by March 31, 1984. This report detailed more than four pages of problems facing ASAT arms control including the following synopsis:

> Deterrence provided by a U.S. ASAT capability would inhibit Soviet attacks against U.S. satellites, but deterrence

is not sufficient to protect U.S. satellites. Because of the potential for covert development of ASAT capabilities and because of the existence of non-specialized weapons which also have ASAT capability, no arms control measures have been identified which can fully protect U.S. satellites. Hence, we must continue to pursue satellite survivability measures to cope with both known and technologically possible, yet undetected, threats. (Executive Office of the President 1984, 9)

ASAT negotiations were restarted in March 1985 as a subset of the broad Defense and Space Talks but were the only area of superpower arms control during the 1980s that did not result in a treaty. Based on resumption of negotiations and continuing administration pressure, congressional restrictions were relaxed and the MHV successfully performed its first and only intercept against a satellite in space on September 13, 1985. By December 1987, following two years of additional funding cutbacks and a congressional prohibition against testing unless the Soviets first performed a dedicated test of their co-orbital system, the Air Force cancelled the MHV. In December 1989, the Army became the executive agent for developing a new ground-launched ASAT system known as the Kinetic Energy ASAT (KE-ASAT); a program DoD formally terminated in 1993 but Congress subsequently provided limited funding of related technologies that have not been tested in space.

The George H. W. Bush Administration continued the Reagan Administration's emphasis on space and issued a number of National Space Policy Directives (NSPDs). This was the first administration since Johnson's to have a formal, standing White House organization, the National Space Council, to coordinate space-policy issues; when Bill Clinton entered office he decided not to staff the Space Council and it has not been used since 1992. In September 1990, Bush continued the policy of launching U.S. government satellites only on U.S. manufactured launch vehicles and in February 1991 in NSPD-3, "U.S. Commercial Space Policy Guidelines," directed U.S. government agencies to use commercial space "products and services to the fullest extent feasible" (Feyock, n.d., 284) to encourage development of this sector, maintain U.S. space preeminence, and save money in areas including: satellite communications, launch services, remote sensing, material processing, and development of commercial infrastructure. NSPD-4, "National Space Launch Strategy," issued on July 10, 1991, supported assured access through a mixed fleet of shuttles and ELVs and directed that DoD and NASA jointly undertake development of a new medium- to heavy-lift vehicle to reduce launch costs while improving reliability and responsiveness. In February and March 1992, President Bush issued NSPD-5 and -6, on Landsat Remote Sensing Strategy and the Space Exploration Initiative; the effort to promote commercial opportunities in Landsat-type remote sensing faced difficulties due to an immature commercial market and the Space Exploration Initiative announced on the 20th anniversary of the *Apollo* 11 landing to return humans to the Moon and then explore Mars was never supported by Congress.

The Clinton Administration was also very active in national space policy. The NSC and the National Science and Technology Council (NSTC) were the lead organizations for developing space policy in the Clinton White House. In September 1993, Vice President Al Gore announced that the Russian Federation would join the International Space Station (ISS) effort and also abide by the terms of the Missile Technology Control Regime (MTCR). In building on the work of the Gore-Chernomyrdin Commission to bring the Russians aboard the ISS, the United States was thinking not just about space cooperation but also had important counterproliferation objectives in employing Russian space scientists as major partners on the ISS and lessening their potential to contribute to the post–Cold War weapons market. The United States paid Russia $400 million in its initial ISS contract, same amount Russia claimed it had forgone by canceling a contract with India for cryogenic rocket engine technology, and provided a total of $800 million in ISS funding to Russia from 1994 to 1998. Unfortunately, however, it is not clear that bringing Russia aboard the ISS appreciably slowed its sale of weapons and dual-use technologies to states of concern such as Iran.

During 1994, Clinton issued policies on Foreign Access to Remote Sensing Space Capabilities in Presidential Decision Directive (PDD)-23 in March, Convergence of U.S.-Polar Orbiting Environmental Satellite Systems in NSTC-2 in May, Landsat Remote Sensing Strategy in NSTC-3 in May, and National Space Transportation Policy in NSTC-4 in August. PDD-23 was designed to encourage U.S. worldwide preeminence in commercial imagery while protecting national security by allowing sales of high-resolution imagery or even entire imagery systems in tightly controlled cases. Such sales can be limited or denied ("shutter control") when the Secretaries of State and Defense determine they might compromise national security or international obligations; the United States has never yet invoked shutter control but has infrequently bought all commercial imagery at certain times over certain areas such as Afghanistan. NSTC-2 directed DoD, the Department of Commerce's National Oceanic and Atmospheric Agency (NOAA), and NASA to work together on a single environmental monitoring system that would meet their requirements and save money. The National Polar-Orbiting Environmental Satellite System (NPOESS) program resulted from this directive; NPOESS repeatedly ran into problems in meeting its requirements on time and on budget resulting in DoD being removed from the program in President Barack Obama's FY 2011 budget. NSTC-3 reflected the failure of Landsat 6 in 1993, transitioned Landsat 7 from DoD to NASA, and established a plan for the U.S. government to continue operating Landsat satellites and maintaining an archive of Landsat-type data. NSTC-4 was an attempt to sustain and revitalize U.S. launch systems by making DoD the lead agency for improving the ELV fleet and NASA the lead agency for developing next generation reusable launch vehicles; the Evolved ELV (EELV) and the X-33 and X-34 programs resulted from this policy. In addition, NSTC-4 made the Departments of Transportation and Commerce responsible for identifying and promoting innovative arrangements to encourage a viable U.S. space-transportation industry and stipulated that the U.S. government would purchase U.S. launch products and services "to the fullest extent feasible" (Feyock, n.d., 331) and not launch government payloads on foreign vehicles except after being granted an exemption by the President.

In 1996, the Clinton Administration issued a new GPS policy as well as its National Space Policy. The fact sheet on NSTC-6, "U.S. Global Position System Policy," was released on March 29 and indicated the United States would continue to provide GPS service without direct user fees, encourage worldwide acceptance and integration of GPS, and discontinue deliberately degrading the accuracy of signals provided to nonmilitary users (Selective Availability) within a decade. Selective Availability was discontinued in May 2000 and in September 2007 the George W. Bush Administration announced that future GPS systems would no longer acquire the capability to degrade signals to nonmilitary users. On September 19, the White House released a fact sheet on PDD-49, "National Space Policy," that reemphasized the United States would maintain a leadership role in space by supporting a strong, stable, and balanced program across all space-activity sectors. Clinton's policy built from the component space policies discussed above and was once again very consistent with previous national space policies stretching back to the Eisenhower Administration. Areas of different or increased emphasis included responsibilities for: NASA and other agencies to develop sensors and acquire data to closely observe and make predictions about Earth's environment; DoD to foster integration and interoperability of satellite control for all governmental space activities; and the Arms Control and Disarmament Agency to identify arms-control issues and opportunities for equitable and effectively verifiable measures that would enhance the security of the United States and its allies.

The George W. Bush Administration issued fact sheets on most of the same space issue areas and again tied them together in a National Space Policy signed on August 31, 2006. The brief recap of these policies below emphasizes areas where Bush's policies diverged from Clinton's. On April 25, 2003, OSTP released a fact sheet on "U.S. Commercial Remote Sensing Space Policy" summarizing National Security Presidential Directive (NSPD)-27, the classified policy which superseded PDD-23, indicating that the U.S. government would: competitively outsource functions in order to rely "to the maximum practical extent" on U.S. commercial remote sensing capabilities for filling its imagery and geospatial needs; focus its remote sensing capabilities on needs that cannot "effectively, reliably, and affordably" be satisfied by commercial providers; and maintain a Sensitive Technology List of "advanced information, systems, technologies, and components" that would be approved for export only rarely, on a case-by-case basis (Office of Science and Technology Policy 2003).

In January 2004, President Bush announced his "Vision for Space Exploration" that sought to advance U.S. scientific, security, and economic interests through a robust space-exploration program including a sustained and affordable human and robotic program to explore the solar system and beyond, starting with a human return to the Moon by the year 2020, in preparation for human exploration of Mars and other destinations. The Constellation program was developed by NASA Administrator Michael Griffin to realize this vision but the Review of U.S. Human Spaceflight Plans Committee (Augustine Committee) reported in October 2009 that implementation would take an additional $3 billion annually and the Obama Administration did not request funding for Constellation in the NASA budget submitted to Congress in February 2010.

On December 15, 2004, a fact sheet on "U.S. Space-Based Position, Navigation, and Timing Policy" was released to explain the policy that superseded NSTC-6 and noted that "the Global Positioning System has grown into a global utility whose multi-use services are integral to U.S. national security, economic growth, transportation safety, and homeland security, and are an essential element of the worldwide economic infrastructure" (Office of Science and Technology Policy 2004). The fundamental goals of this policy include ensuring uninterrupted position, navigation, and timing (PNT) services; remaining preeminent in military PNT; continuing to provide civil services that exceed or are competitive with foreign civil PNT services; remaining an essential component of internationally accepted PNT services; and promoting U.S. technological leadership in PNT applications. In addition, the policy assigned the Secretary of Defense responsibility for development, acquisition, operation, security, and continued modernization of GPS; the Secretary of Transportation lead responsibility for development of requirements for civil applications for the U.S. government; roles for the Secretaries of Commerce, State, Homeland Security, and the Director of National Intelligence to protect the radio spectrum used by GPS, manage international use of GPS, and monitor and respond to domestic and international interference with PNT services; and established a permanent National Space-Based PNT Executive Committee co-chaired by the Deputy Secretaries of the Departments of Defense and Transportation with a Coordination Office to provide secretariat and staff functions for the Executive Committee and an advisory board chartered as a Federal Advisory Committee.

On January 6, 2005, the Bush Administration released a fact sheet on "U.S. Space Transportation Policy" explaining the policy that superseded NSTC-4 including goals to: demonstrate initial capabilities for operationally responsive access to and use of space; develop capabilities for human exploration beyond LEO; and sustain and promote a domestic space- transportation industrial base. The policy also made the Secretary of Defense responsible for maintaining two EELV providers, called for the Secretary to reevaluate this policy with the Director of National Intelligence and NASA Administrator no later than 2010, and charged the NASA Administrator to develop options to meet potential exploration-unique requirements for heavy lift beyond the capabilities of the EELV that emphasize the potential for using EELV derivatives as well as evaluations of the comparative costs and benefits of a new dedicated heavy-lift launch vehicle or options based on the use of shuttle-derived systems.

The final space policy of the Bush Administration was NSPD-49, "U.S. National Space Policy," signed on August 31, 2006, and an unclassified fact sheet about this policy was released on October 6. Bush's policy took a slightly more competitive tone than the Clinton National Space Policy 10 years earlier, indicating "those who effectively utilize space will enjoy added prosperity and security and will hold a substantial advantage over those who do not. Freedom of action in space is as important to the United States as air power and sea power" (Office of Science and Technology Policy 2006). The statement also emphasized that the United States "considers space capabilities—including the ground and space segments and supporting links—vital to its national interests" and "will oppose the development of new legal regimes or other restrictions that seek to prohibit or limit U.S. access to or use of space" (Office of Science and Technology Policy 2006). To achieve the goals of the policy the U.S. government sought to: develop space professionals, improve space system development and procurement, increase and strengthen interagency partnerships, and strengthen and maintain the U.S. space-related science, technology, and industrial base. In addition, the policy made the Secretary of Defense responsible for providing space situational awareness (SSA) for the U.S. government and U.S. commercial space capabilities and services that support national security.

References

Baucom, D. R. 1992. *The Origins of SDI: 1944–1983*. Lawrence: University Press of Kansas.

Burrows, W. E. 1986. *Deep Black: Space Espionage and National Security.* New York: Berkley Books.

Divine, R. A. 1993. *The Sputnik Challenge*. New York: Oxford University Press.

Emme, E. M. 1959. *The Impact of Air Power: National Security and World Politics*. Princeton, NJ: D. Van Nostrand.

Executive Office of the President. March 31, 1984. "Report to Congress: U.S. Policy on ASAT Arms Control."

Feyock, S. (n.d.) "National Security Space Project: Presidential Decisions: NSC Documents." Washington, DC: George C. Marshall Institute.

Hall, R. C. 1995. "Origins of U.S. Space Policy: Eisenhower, Open Skies, and Freedom of Space." In *Exploring the Unknown: Selected Documents in the History of the U.S. Civil Space Program*, Vol. I: *Organizing for Exploration*, ed. John M. Logsdon, 213–229. Washington, DC: NASA History Office.

McDougall, W. A. 1985. . . . *the Heavens and the Earth: A Political History of the Space Age*. New York: Basic Books.

National Security Decision Directive Number 42. "National Space Policy." July 4, 1982. 2–3, NSC box, National Archives, Washington, D.C.

Nitze, P. H. February 20, 1985. "On the Road to a More Stable Peace." Bureau of Public Affairs, Department of State, Current Policy No. 657.

Office of Science and Technology Policy. May 13, 2003. "Fact Sheet: U.S. Commercial Remote Sensing Space Policy." Washington, DC: The White House.

Office of Science and Technology Policy. December 15, 2004. "Fact Sheet: U.S. Space-Based Positioning, Navigation, and Timing Policy." Washington, DC: The White House. Office of Science and Technology Policy. October 6, 2006. "U.S. National Space Policy." Washington, DC: The White House.

Schoettle, E. C. B. 1966. "The Establishment of NASA." In *Knowledge and Power: Essays on Science and Government*, ed. Sanford A. Lakoff. New York: Free Press.

Spires, D. N. 2007. *Beyond Horizons: A History of the Air Force in Space, 1947–2007*. Colorado Springs, CO: Air Force Space Command History Office.

Stares, P. B. 1985. *The Militarization of Space: U.S. Policy, 1945–1984*. Ithaca, NY: Cornell University Press.

Technological Capabilities Panel of the Science Advisory Committee. 1955. *Meeting the Threat of Surprise Attack, Vol. II* (Washington, D.C., February 14); cited in R. Cargill Hall, "Sputnik, Eisenhower, and the Formation of the U.S. Space Program," *Quest* vol. 14 no. 4 (2007): 32–39.

U.S. Congress, House, Committee on Government Operations. 1959. *Organization and Management of Missile Programs: Hearings before the Committee on Government Operations*, 86th Cong., 1st sess. Cited in Michael H. Armacost, *The Politics of Weapons Innovation: The Thor-Jupiter Controversy* (New York: Columbia University Press, 1969), 227.

U.S. Congress, House, Committee on Science and Astronautics. 1961. Memorandum, "Coordination of Satellite and Space Vehicle Operations" in *Defense Space Interests: Hearing before the Committee on Science and Astronautics*, 87th Cong., 1st sess.

U.S. Congress, Senate, Committee on Foreign Relations. 1967. *Treaty on Outer Space: Hearing before the Committee on Foreign Relations*, 90th Cong, 1st sess.

U.S. Department of State. 1958–1960. *Foreign Relations of the United States, 1958–1960, Vol. II.* Washington, DC: GPO.

U.S. Department of State. 1990a. *Foreign Relations of the United States, 1955–1957, Vol. XIX: National Security Policy.* Washington, D.C.: GPO.

U.S. Department of State. 1990b. *Foreign Relations of the United States, 1955–1957, Vol. XI.* Washington, DC: GPO.

Notes

1. McDougall's description is based on the Goodpaster notes at the Eisenhower Library. On this meeting see also Robert A. Divine, *The Sputnik Challenge* (New York: Oxford University Press, 1993), 5–7.

2. Directive 5160.32 and the news release accompanying its release are reprinted in House, *Defense Space Interests*, 2–4.

3. Two complete pages and approximately five additional paragraphs are deleted from the sanitized version of this directive. The White House also issued a five-page fact sheet titled "National Space Policy," on July 4, 1982, reprinted in National Aeronautics and Space Administration, *Aeronautics and Space Report of the President, 1982 Activities* (Washington, DC: GPO, 1983), 98–100.

2

Problems, Controversies, and Solutions

Chapter 1 revealed many enduring issues and tensions in U.S. space policy both between and within each of the four sectors of space activity including: the proper level of resources and emphasis for exploration, scientific, and applications space missions as well as whether these missions are best conducted by humans or robots; the appropriate role, if any, for human spaceflight in supporting intelligence and military missions; the proper level of space militarization or potentially weaponization and the prospects for offensive and defensive missions in space; space's role in enabling transformed terrestrial military capabilities and the use of space capabilities for ballistic missile defense or other force application missions; the proper balance between public and commercial space activity as well as the most appropriate mechanisms to stimulate growth in commercial space activity; and the best mix of unilateral or cooperative space ventures and appropriate ways to educate and develop space professionals, and protect or share space technology via export controls and other mechanisms. This chapter explores in greater depth three issue areas that relate primarily to domestic politics: space and military transformation, space regulation and export controls, and space and missile defense.

Space and Military Transformation

Strategic analysts debate whether modern technology can change the basic character of warfare and how much it has modified fundamental precepts such as mass and the fog of war, but most agree

47

that modern technologies including space capabilities have radically altered the tactics and conduct of war. The evolution of warfare through World Wars I and II created a fearsome attrition-based war machine by coupling the increasingly lethal products of the industrial revolution with improved military organizations and doctrine. Modern attrition warfare also necessitated development of what Stephen Biddle calls the modern system: a complex combined arms approach to fire, maneuver, and concealment that enables survival and military effectiveness but requires an adaptive and well-trained military to produce the skills required for success in the modern battle space. The modern system exacerbates disparities in military effectiveness because militaries that lack the resources to adopt a complex combined arms approach or fail to adapt are punished severely in conventional warfare (Biddle 2004). These factors have also created incentives for development of "hybrid warfare," an approach that attempts to exploit sanctuaries associated with traditional legal constructs for warfare and other vulnerabilities by employing all forms of war and tactics (conventional, irregular, and terrorist), perhaps even simultaneously.

The national security space (NSS) sector includes DoD activities, conducted primarily by the Air Force, to enhance national security and National Reconnaissance Office (NRO) programs to collect intelligence data from space. The NSS sector is also divided sometimes into separate sectors known as the military or defense space sector and the intelligence space sector. Two major analytical frameworks shape many discussions about NSS capabilities and how they fit into the modern system and hybrid warfare: military space mission areas and military space doctrines. There are four military space mission areas: space support, force enhancement, space control, and force application. Currently, force enhancement is the most important military space mission area; due to growth in the efficacy of space capabilities, many analysts believe they now have moved beyond just enhancing the effectiveness of terrestrial forces and enable a wider range of military missions to be undertaken or even contemplated. Table 2.1 shows the major divisions within force enhancement as well as the current and projected space systems that support these missions (systems listed below the dotted line have not yet been deployed and those in italics have been recently canceled). Building on the analysis in David Lupton's 1988 book, *On Space Warfare*, there are also four major military space doctrines: sanctuary, survivability, control, and high ground (Lupton 1988). The attributes associated with these doctrines—primary value and functions, employment strategies, conflict missions, and desired organizational structures—are shown in Table 2.2.

The United States has been at the forefront of employing the modern system and developing a space-enabled global reconnaissance, long-range precision strike network. Operation Desert Storm in 1991 marked the emergence of space-enabled warfare when a wide range of space systems including those designed for

TABLE 2.1

Force Enhancement Mission Areas, Primary Orbits, and Associated Space Systems

Environmental Monitoring	Communications	Position, Navigation, and Time (PNT)	Integrated Tactical Warning and Attack Assessment	Intelligence, Surveillance, and Reconnaissance (ISR)
Polar Low-Earth Orbit (LEO)	Geostationary-Earth Orbit (GEO) and LEO	Semi-synchronous Orbit	GEO and LEO	Various
Defense Meteorological Support Program (DMSP) ---------------- National Polar-Orbiting Operational Environmental Satellite System (NPOESS)	Defense Satellite Communications System (DSCS) II, DSCS III, Ultra-High Frequency Follow-on (UFO), Milstar, Global Broadcast System (GBS), Iridium, Wideband Global System (WGS), commercial systems ---------------- Advanced Extremely High Frequency (AEHF), Mobile User Objective System (MUOS), Polar Military Satellite Communications System, Transformational Satellite Communications System (TSAT)	Global Positioning System (GPS) GPS II GPS IIR GPS IIR-M ---------------- GPS IIF GPS III	Defense Support Program (DSP), GPS ---------------- Space-Based Infra-Red Systems (SBIRS), Space Tracking and Surveillance System (STSS), Third-Generation Infrared Surveillance Program	Imaging (IMINT) Satellites, Signals Intelligence (SIGINT) Satellites, commercial systems ---------------- Integrated Overhead SIGINT Architecture (IOSA), Future Imagery Architecture (FIA), Space Radar

TABLE 2.2
Attributes of Military Space Doctrines

	Primary Value and Functions of Military Space Forces	Space System Characteristics and Employment Strategies	Conflict Missions of Space Forces	Appropriate Military Organization for Operations and Advocacy
Sanctuary	• Enhance Strategic Stability • Facilitate Arms Control	• Limited Numbers • Fragile Systems • Vulnerable Orbits • Optimized for National Technical Means (NTM) verification mission	• Limited	National Reconnaissance Office (NRO)
Survivability	Above functions plus: • Force Enhancement	• Terrestrial Backups • Distributed Architectures • Autonomous Control	• Force Enhancement • Degrade Gracefully	Major Command or Unified Command
Control	• Control Space • Significant Force Enhancement	• Hardening • On-Orbit Spares • Crosslinks • Maneuver • Less Vulnerable Orbits • Stealth • Attack Warning Sensors	• Control Space • Significant Force Enhancement • Surveillance, Offensive, and Defensive Counterspace	Unified Command or Space Force
High Ground	Above functions plus: • Decisive Impact on Terrestrial Conflict • Ballistic Missile Defense (BMD)	• 5 Ds: Deception, Disruption, Denial, Degradation, Destruction • Reconstitution Capability • Active Defense • Convoy	Above functions plus: • Decisive Space-to-Space and Space-to-Earth Force Application • BMD	Space Force

Cold War strategic missions such as the Defense Support Program (DSP) missile launch detection satellites and other constellations that were not yet completed such as the Global Positioning System (GPS) produced transformational effects from the lowest tactical level, for instance guiding individual vehicles across trackless deserts, up through the highest strategic level, including helping to keep Israel out of the conflict. In Operation Desert Storm, less than 8 percent of air-delivered munitions were precision-guided (none by GPS) and satellites provided only one megabit per second (Mbps) communications connectivity to battalion-sized units deployed in theater; by the time of Operations Enduring Freedom and Iraqi Freedom in 2001 and 2003, almost 70 percent of air-delivered munitions were precision-guided (mostly by GPS) and satellites provided communications at speeds over 50 Mbps to deployed battalions. During Operation Desert Storm, all coalition aircraft operations were controlled by an inflexible air tasking order (ATO) that took 72 hours to develop; moreover, because aircraft carriers and other deployed locations were unable to receive the ATO electronically, a printout had to be manually delivered every cycle. In stark contrast, during Operations Enduring Freedom and Iraqi Freedom, the majority of aircraft took off on their combat sorties without assigned targets; they were dynamically tasked in flight onto targets that emerged during their sortie or attacked remaining targets assessed as most important after their arrival on station.

This acceleration of space-enabled capabilities today allows U.S. commanders to draw from worldwide intelligence, surveillance, and reconnaissance (ISR) and analysis, communicate faster, strike more accurately, and assess operational effectiveness in real time. Space capabilities have become so seamlessly integrated into the overall U.S. military structure that commanders can remain focused on strategic objectives instead of making tactical decisions on how to engage specific targets. The United States continues to develop lighter and more easily deployable forces that are better able to leverage space and network-enabled operations and strike more precisely from greater distances to achieve full spectrum dominance over adversaries that may range from emerging military peers to insurgents and terrorists. Space-enabled warfare can deliver highly precise effects, minimize collateral damage, and shorten the duration of conflict, but should be part of a balanced portfolio of capabilities that encourages pursuit of political objectives using all appropriate tools of statecraft and reduces temptations to overuse or inappropriately use the military instrument of power.

Space capabilities often provide the best and sometimes the only way to pursue America's ambitious defense transformation goals. There are, however, many difficult and fundamental issues related to space and defense policy including: the role of space capabilities in enabling the information revolution and the new American way of war; changes caused by growth in commercial space activity and the number of major space actors; and the role and efficacy of space capabilities in structuring options for military intervention as well as in dissuading and deterring competition from potential adversaries in the changed geopolitical environment following the end of the Cold War, the 9/11 attacks, and Operations Enduring and Iraqi Freedom. These complex factors contribute to uncertainty about how space capabilities can best advance U.S. national security, the most useful organizational structures to manage and transform space activities themselves, and the utility of investments in space capabilities versus other enabling military capabilities. Moreover, the United States faces significant challenges in its current plans to modernize, improve, or replace almost all major national security space systems because they are required for future transformed forces but their acquisition has been marked by cost overruns and deployment delays. It is unclear whether the United States will be able to find and follow the best path forward for space strategy, implement the best management and organizational structures for space activities, and sustain the political will needed to continue funding the nearly simultaneous modernizations currently planned. It is also uncertain whether these new and improved space capabilities can be developed and integrated on cost and on time and whether these future systems will deliver on their promise of continuing to accelerate transformational capabilities and effects.

Following implementation of one of the recommendations of the January 2001 Commission to Assess National Security Space Management and Organization (Space Commission) Report, DoD used an accounting procedure known as the virtual major force program (vMFP) for NSS spending, but it has remained difficult to track overall and acquisition funding for NSS for several reasons including a lack of consistency with respect to the specific programs that were included in the vMFP in any given year and the classified funding for many NSS programs. Congress has repeatedly called for creation of a "hard" MFP (MFP-12) for space and designation of an Office of the Secretary of Defense (OSD) official to provide overall supervision of the preparation and justification of program recommendations and budget proposals included in MFP-12, but DoD has been slow to institute this approach. According to the very broad definition of "space" funding in the Space Foundation's *Space Report*, total NSS funding amounted to $43.53 billion in Fiscal Year (FY) 2009 (DoD $26.53 billion, NRO $15 billion, and National Geospatial-Intelligence Agency (NGA) $2 billion). Under a far more constrained definition of space funding, former Representative Ellen Tauscher, chairwoman of the House Strategic Force Subcommittee, estimated DoD space spending in the FY 10 authorization to be $11 billion. Regardless how NSS funding is tracked, it is clear that the currently planned path ahead for significant improvements and modernizations to almost all NSS systems will be very difficult, if not unsustainable, especially if current programs follow the historic trend of an average 69 percent rise in costs for space research, development, engineering, and testing as well as an average growth of 19 percent in space procurement costs.

The remaining discussion in this section focuses on the most significant ongoing modernizations in NSS capabilities. Currently, the Air Force maintains a constellation of infrared sensors on GEO satellites, called the Defense Support Program (DSP), to provide warning of ballistic missile launches worldwide and some data on the type of attack and the missile's intended target. The last DSP satellite (DSP-23) was successfully launched from Cape Canaveral aboard a Delta IV-Heavy Evolved Expendable Launch Vehicle (EELV) in November 2007 but failed after reaching GEO. The follow-on to DSP is the Space-Based Infrared Systems (SBIRS), a program designed to support overhead persistent infrared (OPIR) needs in four mission areas: missile warning, missile defense, technical intelligence, and battlespace awareness. Lockheed Martin is the prime contractor and Northrop Grumman builds the staring and scanning infrared sensors for SBIRS. The operational SBIRS constellation will consist of GEO satellites and sensors on classified highly elliptical orbit (HEO) satellites. The HEO sensors are operational and have been certified for missile warning and technical intelligence applications; the first GEO satellite is now scheduled to launch in FY 11. In September 2009, the Missile Defense Agency (MDA) launched two demonstration satellites from a Northrop Grumman system formerly known as SBIRS-Low and now named the Space Tracking and Surveillance System (STSS). The demonstrators are expected to be available from two to four years and test key areas including birth-to-death tracking of strategic and tactical missiles, integrated tracking and timely handoffs to defense systems, and risk reduction for future satellite development. Work has also begun on the Precision Tracking Space System, a next-generation space sensor layer designed to track every missile accurately and transmit fire-control quality data to defense systems. Unfortunately, SBIRS has been a troubled NSS acquisition effort; a Defense Science Board report called it "a case study for how not to execute a space program" (Defense Science Board 2003, 6), total cost estimates are almost five times higher and deployment many years behind schedule; and the program has triggered four required reports to Congress under the Nunn-McCurdy Act. The program has had fewer problems and been on a much better trajectory for success after being fundamentally restructured in December 2005. Fundamental U.S. goals for position, navigation, and timing (PNT) include: maintaining uninterrupted PNT services for all user needs; remaining preeminent in military PNT; providing civil services that exceed or are competitive with foreign PNT services and continue as an essential component of internationally accepted PNT services; and promoting U.S. leadership in PNT. One of the most difficult challenges is mandated by Navigation Warfare requirements to: operate GPS effectively despite adversary jamming; deny use to adversaries; not unduly disrupt civil, commercial, or scientific uses outside an area of military operations; and identify, locate, and mitigate interference on a global basis. The Air Force acquires and operates the GPS constellation that currently contains 32 satellites developed through a series of block upgrades. In September 2005, the Air Force began launching Lockheed Martin block IIR-M satellites, which incorporate two new military signals and a second civilian signal. The first Boeing block IIF satellite, which broadcasts a third signal for civilian use, was launched in May 2010. The first block III satellites (GPS IIIA), currently scheduled for first launch in 2014, will transmit a new civilian signal designed to be highly interoperable with the European Galileo and Japanese Quazi-Zenith Satellite Systems as well as improving the anti-jam capability of M-Code signals for military users.

On February 1, 2010, the Office of Science and Technology Policy at the White House announced it will end a troubled effort to develop a single civil-military weather satellite system, the National Polar-orbiting Operational Environmental Satellite System (NPOESS), and instead pursue two separate lines of polar-orbiting satellites to serve military and civilian users. The NPOESS program arose from a 1994 decision (NSTC-2) by the Clinton Administration to converge the polar-orbiting weather satellite systems traditionally built and operated separately by NOAA and the Air Force. The converged approach was projected to save money, but

NPOESS wound up being more technologically challenging than anticipated, was delayed, and overran original cost estimates by several billion dollars. Under the new plan, a precursor satellite, the NPOESS Preparatory Project remains on track for a September 2011 launch; the Air Force is scheduled to launch its two remaining Defense Meteorological Satellite Program (DMSPs) in 2012 and 2014; NOAA has requested $1.06 billion in 2011 and an additional $3.47 billion over next the four years for a new system, the Joint Polar Satellite System (JPSS), that will be managed by NASA's Goddard Space Flight Center and is scheduled for first launch in 2015, and the Air Force plans to develop a new Defense Weather Satellite System (DWSS).

As its forces have become increasingly expeditionary and network enabled, DoD's reliance on commercial communications satellites (comsats) has grown sharply. During Operations Iraqi and Enduring Freedom over 80 percent of telecommunications with deployed forces in the theaters of operation were carried over commercial comsats. Benefits of using commercial systems include lower acquisition and operations costs as well as greater flexibility but must be balanced against drawbacks such as high cost of buying commercial services on the spot market, competition for available bandwidth, and less secure and protected systems. DoD is developing and beginning to deploy a number of more capable, dedicated comsat systems including the Wideband Global System (WGS), the Advanced Extremely High Frequency (AEHF), and the Transformational Satellite Communications System (TSAT). The first WGS launch had been delayed for over three years and was accomplished on October 10, 2007; the six-satellite constellation is due to be completed in 2013 and remain in service until 2024. On October 4, 2007, Australia announced it will pay $823.6 million to purchase WGS-6 as a way to enhance the network-enabled capabilities of Australian forces, leverage the entire WGS constellation, and further the close ties that already exist between the United States and Australia. Protected communications capabilities are currently provided by five Milstar satellites that are expected to be operational at least until 2014. Beginning with the first launch in August 2010, a constellation of five or perhaps six Advanced EHF (AEHF) will begin replacing Milstar. AEHF includes cryptography necessary to provide worldwide, survivable, and anti-jam protected communications as well as much higher data rates than Milstar. Lockheed Martin is the prime contractor on AEHF, Northrop Grumman is developing the satellite payload and it is a cooperative program with Canada, the United Kingdom, and the Netherlands. The most important planned future system for both protected and wideband communications capabilities had been TSAT, a system originally scheduled for first launch in 2013, extensively restructured and delayed until 2019, and then cancelled in 2009. TSAT was envisioned as a five-satellite system with a laser crosslink ring in GEO providing DoD with both high-data-rate wideband and protected communications and enabling communications- and networking-on-the-move by connecting thousands of users simultaneously through networks rather than using limited point-to-point circuits. Currently, it is not clear how DoD will acquire the wideband and especially protected capabilities required for its future force structure.

Many components of the U.S. ISR network are classified but at least portions of the Future Imagery Architecture (FIA), Space Radar (SR), and commercial remote sensing systems are public knowledge. In September 1999, NRO rocked the aerospace industry by selecting Boeing to build the FIA, its next generation imaging satellites, bypassing Lockheed Martin, the decades-long incumbent on the program. The original design for the FIA electro-optical (EO) and radar-imaging satellites called for a constellation that split collection functions among smaller, simpler, and more numerous satellites in order to collect more imagery with more frequent revisit rates. FIA, however, soon stumbled badly due to a host of problems plaguing most government satellite programs during this period that included overly optimistic initial bids, unrealistic cost caps and lack of management reserves, and granting total systems performance responsibility (TSPR) to contractors while government oversight capabilities and responsibilities languished. By 2005 the price had grown from $6 billion to as much as $18 billion, prompting Director of National Intelligence (DNI) John Negroponte to cancel the program. Boeing will continue developing radar-imaging satellites but, at the DNI's direction, the EO portion of the architecture was downscaled and reassigned to Lockheed Martin. The Space Radar program was a joint program being designed to satisfy both national intelligence and joint warfighter requirements for a global capability to detect, image, and track mobile targets in denied areas during all weather conditions. This unclassified program was restructured then terminated in 2008.

Space-control capabilities are key enablers of all NSS activity. These programs focus on developing ground- and space-based sensors to enhance space situational awareness or SSA (knowledge of activity and events in or that could affect circumterrestrial space), improve capabilities to protect friendly space assets from enemy attack, and develop capabilities to negate enemy space capabilities. A very important addition to U.S. SSA capabilities launched in August 2010: The Space Based Surveillance System is designed to have significantly improved capabilities compared with previous space-based sensors, including twice the sensitivity and speed at detecting objects, 3 times better probability of detecting events, and 10 times greater capacity to observe events. Joint Publication 3–14, *Joint Doctrine for Space Operations*, discusses ways to gain or maintain space control by providing freedom of action through protection and surveillance or to deny freedom of action through prevention and negation. Air Force doctrine, by contrast, aligns space-control doctrine like air doctrine as offensive counterspace (OCS) and defensive counterspace (DCS). OCS missions would disrupt, deny, degrade, or destroy space systems, or the information they provide, if used for purposes hostile to U.S. national security interests. DCS missions include both active and passive measures to protect U.S. and friendly space-related capabilities from enemy attack, interference, or use for purposes hostile to U.S. national security interests. As reflected in NSPD-49, repeated congressional language, and the January 2007 Chinese anti-satellite (ASAT) test, space control has become an area of increasing concern, attention, and funding. Potential development of systems with the ability to apply force to, in, or especially from, space is of even greater congressional, public, and international concern. These concerns are exacerbated by significant difficulties in distinguishing between concepts and technologies being developed for ballistic missile defense, protection, space control, and force application as well as the development of some of these systems in classified programs.

Space Regulation and Export Controls

It is not clear that the space regulations and exports controls developed mainly during the Cold War are advancing U.S. economic and security interests, especially as space becomes an increasingly competitive, congested, and perhaps even contested environment. To date the only space activity that has consistently generated wealth are communications satellites (comsats). Comsats, in turn, form just a small segment of the complex and evolving global telecommunications markets and regulatory regimes. Recent growth in the commercial space sector has largely been driven by more use of space-enabled telecommunications services and by direct-to-home services in particular. The global comsat market is also fundamentally shaped by technological considerations such as the competition between fiber optic cables and comsats. Over time, the primary debates over shaping telecommunication markets have moved from the North-North and East-West issues of the 1960s, to the North-South issues of the 1970s and 1980s, and have now returned to North-North economic competition issues. These trends also carry increasing implications for national security as the military becomes more reliant on comsats and other commercial space services.

In contrast to the largely informal mechanisms that evolved to legitimize spysats, the United States was also the primary driver behind the creation of several important formalized arms-control and regulatory regimes for space. In the first of these, the regulatory regime for comsats, Washington moved to structure the commercial environment of space and set several precedents for commercial space that are still in place today. The Communications Satellite Act of 1962 was the first step, creating the Communications Satellite Corporation (Comsat) as a government chartered company initially equally shared between private stockholders and existing public common-carrier telecommunication corporations. In addition, the regime was structured to advance U.S. foreign policy and national security goals, a way for "all nations to participate in a communications satellite system, in the interests of world peace and a close brotherhood among peoples throughout the world." The act envisioned three basic criteria for an international system: (1) a single global system open to participation by any state in the world; (2) a commercial enterprise; and (3) a system designed to provide service to both the developed and developing states of the world. By August 1964, delegates from 14 states were able to hammer out the Interim Agreement that marked the unofficial birth of the International Telecommunications Satellite Consortium (Intelsat) then 84 states signed the Definitive Agreement in August 1971.

By the early 1980s, significant controversies again swirled around the organization's purpose and focus. President Ronald Reagan's Administration was eager to reap benefits it believed would follow from breaking up government regulated domestic monopolies such as the American Telephone and Telegraph (AT&T) Corporation and also favored removing protections from international monopolies such as Intelsat. Presidential Determination 85–2 of November 1984 found "that international satellite systems separate from Intelsat are required in the national interest" but this finding was balanced by the 1985 separate system policy (SSP). The SPP prohibited Intelsat competitors from accessing the public switched networks (i.e., the public telephone system) to serve U.S. customers and was a formidable entry constraint because it protected Intelsat from competition on services that accounted for almost three-quarters of its revenues.

This opening for limited comsat competition began to erode Intelsat's position in the markets for every service it provides (voice, data, video, and audio). But it was the emergence of a very strong transoceanic fiber optic cable infrastructure that began in the late 1980s and accelerated throughout the 1990s that ended Intelsat's dominance. By the late 1990s, through the Open-market Reorganization for the Betterment of International Telecommunications (ORBIT) Act, Congress moved U.S. comsat policy full circle by amending the Communications Satellite Act of 1962 to create incentives for competition and privatization. More specifically, the ORBIT Act called for Intelsat to be privatized in a pro-competitive manner no later than April 1, 2001, and prohibited the Federal Communications Commission (FCC) from issuing or renewing licenses to Intelsat or its privatized successor organization unless they would not harm competition in the U.S. telecommunication market. The privatization of INTELSAT on July 18, 2001, marked a complete reordering of the regulatory regime for comsats as originally designed in the early 1960s.

Comsats are regulated by a number of international and domestic organizations including the International Telecommunications Union (ITU), Federal Communications Commission (FCC), and National Telecommunications and Information Agency (NTIA). The ITU operates under a "federal" permanent structure but its most important decisions for regulating the radio spectrum are negotiated at the World Radiocommunication Conferences (WRCs) now normally held about every four years. The ITU has and will continue to play a primary role in structuring the place of comsats in the world telecommunications regime; obtaining from the ITU the radio frequency license required to operate a comsat can be a difficult and time-consuming process. Unlike Intelsat, however, it is not likely this specialized UN agency will be a candidate for privatization, primarily because it serves in a more purely regulatory fashion, licenses a limited resource, and provides a global public good by reducing harmful interference between transmissions throughout the radio spectrum. Similar to Intelsat, the primary axes of controversy over ITU regulation have shifted from North-North and East-West issues in the 1960s to North-South issues during the 1970s and 1980s and recently have returned primarily to North-North issues of economic competitiveness. The WRCs held in 1985 and 1988 developed an Allotment Plan and Improved and Simplified Procedures for the ITU satellite coordination/registration process that were important in helping to resolve North-South conflict caused by increasing use of GEO and reduced fears in the developing world about being blocked from access to this limited resource. Conflicts during the past several WRCs have increasingly centered on controversies associated with the perceived need for new and expanded spectrum for commercial applications, disagreements over communications protocols and standards, and the targeting of different current services as potential sources of additional commercial spectrum.

The FCC and the Department of Commerce's NTIA are the final major players in the comsat regulatory regime. The FCC has jurisdiction over the entire frequency spectrum except for the portion used by the federal government that is managed by the NTIA. These organizations are the United States' domestic regulatory counterparts to the ITU and they work in conjunction with the Department of State to regulate and represent the commercial and diplomatic interests of the United States and its comsat industry. The FCC is an independent United States government agency that was established by the Communications Act of 1934 and is charged with regulating interstate and international communications by radio, television, wire, satellite and cable. NTIA is the President's principal adviser on telecommunications and information policy issues. It was created in 1978 by merging the White House Office of Telecommunication Policy with the Commerce Department's Office of Telecommunications. One of the greatest distinctions between the approach of the FCC and the ITU in granting spectrum allocations is that since 1994, the FCC used auctions to award licenses to use some parts of the spectrum to the highest bidders. Although there are significant economic and legal issues, many analysts believe that the FCC's use of auctions to allocate scarce radio frequency resources can be a more economically rational system for allocating spectrum, generate income for the government, and help reduce the time for processing a license. Use of auctions to allocate comsat spectrum by the FCC and other domestic regulators may put increasing pressure on the ITU to consider the use of auctions to allocate comsat spectrum internationally.

High-resolution commercial remote sensing is an evolving and complex issue area that requires carefully considered regulation and transparency- and confidence-building measures (TCBMs) to help balance several interdependent goals. High-resolution commercial imagery creates opportunities and risks across a wide range of diplomatic, military, economic, and political considerations including: how these systems contribute to global transparency and the implications of a more transparent world; economic competition and the viability of the high-resolution commercial remote sensing industry worldwide; competition from other remote sensing providers and the quality, timeliness, and types of products offered by space-based systems; the optimal balance between commercial and government systems; and mechanisms for controlling and regulating this industry.

Following the end of the Cold War, the United States completely reoriented its policy on high-resolution commercial remote sensing away from the secret spysat regime crafted by the Eisenhower Administration at the opening of the space age. Under the Land Remote Sensing Policy Act of 1992 and National Security Presidential Directive (NSPD)-27 of 2003, it is now the policy of the United States to create incentives to develop a high-resolution commercial remote sensing industry. By attempting to dominate this market worldwide, the United States hopes to preserve its industrial base, leverage commercial systems for government uses, and shape global standards on acceptable use via mechanisms such as shutter control.

NSPD-27 is designed to balance economic considerations with national security concerns. It details a number of specific restrictions on high-resolution commercial imagery products and systems as well as U.S. government responsibilities to: competitively outsource functions in order to rely "to the maximum practical extent" on U.S. commercial remote sensing capabilities for filling its imagery and geospatial needs; focus its remote sensing capabilities on needs that cannot "effectively, reliably, and affordably" be satisfied by commercial providers; and maintain a Sensitive Technology List of "advanced information, systems, technologies, and components" that would be approved for export only rarely, on a case-by-case basis (Office Science and Technology Policy 2003).

Like most dual-use technologies, high-resolution commercial remote sensing holds both beneficial and threatening potential. By increasing transparency, commercial imagery should help to dampen the security dilemma by illuminating the actual force levels of states and increase stability by revealing preparations for an attack. Conversely, by pinpointing potential targets, such systems may create incentives for preemption—especially by states that possess highly accurate, long-range weapons. Paradoxically, the amount of data available from proliferated commercial imagery systems might place greater value on the type of information usually only available from human intelligence sources—the intentions of potential adversaries. In addition, this technology can empower non-state actors and provide them with information to support a wide variety of goals. Moreover, digitized imagery is also ideally suited for deception because the raw data must be mathematically processed to create images and this processing can be manipulated in a variety of ways, including techniques not available for altering film and that even experts can have difficulty detecting. Proliferation of commercial satellite imagery creates pressures for development of more robust denial, deception, and ASAT countermeasures; raises legal conundrums and questions about the admissibility of such imagery in court; and increases responsibilities for governments and the media to confirm the veracity of satellite imagery.

The U.S. government should continue to study and evaluate the evolution of the commercial remote sensing market to ensure that its policy objectives are being met. Regulatory mechanisms such as shutter control that the United States has put in place appear to provide an equitable balance between economic considerations and national security concerns. These mechanisms should also be self-regulating to a large degree. If the United States overuses shutter control it may drive potential customers to foreign imagery providers; but such a control seems prudent before the United States creates incentives for its high-resolution commercial remote sensing industry to dominate the global market. This area also offers potential for novel means of control and exploitation. The requirement for imagery providers to use only approved encryption devices that allow U.S. government access during periods of shutter control, especially when coupled with the potential to use digital data for deception, certainly presents some interesting possibilities for control and exploitation by leaving systems operating rather than shutting them off. Finally, the U.S.

government needs to do a better job in ensuring that all agencies are using commercial remote sensing products to the maximum practical extent; focusing their remote sensing systems on meeting needs that cannot be met by commercial providers; and, most importantly, holding accountable organizations not complying with this policy guidance.

Because of the growing and ubiquitous importance of space capabilities, some analysts believe they should now be considered in a new way as global utilities that provide an essential foundation for globalization, the global economy, and the information age. Global utilities can be defined as: civil, military, or commercial systems—some or all of which are based in space—that provide communication, environmental, position, image, location, timing, or other vital technical services or data to global users. The United States is the world's leading provider of global utilities and supplies environmental monitoring data and GPS signals free to all users worldwide as a public good. To date, all space-based global utilities provide information services but they are analogous to Earth-bound utility services that provide a foundation for modern life such as water and electricity. And like these Earth-bound utility services, space-based global utilities may be subject to regulation and control at the local, state, national, and international levels. Two relatively minor failures illustrate just how embedded global utilities have become in the modern networked world: in 1996, a controller at the Air Force GPS control center accidentally put the wrong time into just one of the satellites and this erroneous signal was broadcast for just six seconds before automatic systems turned the signal off. That momentary error caused more than 100 of the 800 cellular telephone networks on the U.S. East Coast to shut down and some took hours or even days to recover. In May 1998, 40–45 million pager subscribers lost service; some automatic teller machines and credit card machines could not process transactions; news bureaus could not transmit information; and many areas lost television service—all because of the failure of *one* comsat, the Galaxy 4.

How global utilities should be controlled, regulated, and protected are complex issues that are shaped by a number of factors including the specific capabilities in question, the services they provide, and the primary users. Given the current range of existing regulation and control for global utility services, it is not clear what national security or economic objectives would be served by attempting to regulate these services in the same or even similar ways. Protection of global utilities and other space capabilities presents daunting challenges. Despite the growing number of threats and incidents of unintentional and purposeful interference, to date there has been almost no demand from the operators of commercial communications satellites for defense of their multibillion dollar assets. The current lack of support from industry for protection of global utilities is particularly disappointing to those who began advocating in the late 1990s that the "flag follows trade," advancing the argument that protection is needed and will be increasingly demanded as space commercialization grows. Other analysts believe that a multilateral approach to protection for global utilities would be best and argue that this function should be advanced through TCBMs developed through an international organization such as the UN. This approach would likely, however, be filled with all the political, bureaucratic, economic, and technical difficulties that have plagued almost all international space efforts. The long and rocky path to complete the International Space Station does not inspire great confidence in this approach to providing protection for global utilities. At the opposite end of the spectrum are those who advocate that DoD, and the Air Force in particular, should take on the global utility protection mission regardless of international opposition or a lack of support from industry. In addition to the likely domestic and international political opposition to this approach, creating effective assurances and protections for space capabilities also faces difficult economic and especially technical challenges.

History suggests there is a very important role for militaries both in setting the stage for the emergence of international legal regimes and in enforcing the norms of those regimes once they emerge. Development of any TCBMs for space, such as rules of the road or codes of conduct, should draw closely from the development and operation of such measures in other domains such as sea or air. The international community should consider the most appropriate means of separating military activities from civil and commercial activities in the building of these measures because advocating a single standard for how all space activities ought to be regulated or controlled is inappropriately ambitious and not likely to be helpful. The U.S. Department of Defense requires safe and responsible operations by warships and military aircraft but they are not legally required to follow all the same rules as commercial traffic and sometimes operate within specially protected zones that separate them from other traffic. Full and open dialogue about these ideas along with others will help develop space rules that draw from years of experience in operating in these other domains and make the most sense for the unique operational characteristics of space. Other concerns surround the implications of various organizational structures and rules of engagement for potential military operations in space. Should such forces operate under national or only international authority, who should decide when certain activities constitute a threat, and how should such forces be authorized to engage threats, especially if such engagements might create other threats or potentially cause harm to humans or space systems? Clearly, these and a number of other questions are very difficult to address and require careful international vetting well before actual operation of such forces in space. Finally, consider the historic role of the Royal and U.S. Navies in fighting piracy, promoting free trade, and enforcing global norms against slave trading. Should there be analogous roles in space for the U.S. military and other military forces today and in the future? What would be the space component of the Proliferation Security Initiative and how might the United States and others encourage like-minded actors to cooperate on such an initiative? Attempts to create legal regimes or enforcement norms that do not specifically include and build upon military capabilities are likely to be divorced from pragmatic realities and ultimately frustrating efforts.

Seemingly new United States focus and direction on space TCBMs initially was provided by a statement on the Obama Administration White House Web site that appeared on January 20, 2009: "Ensure Freedom of Space: The Obama-Biden Administration will restore American leadership on space issues, seeking a worldwide ban on weapons that interfere with military and commercial satellites." The language about seeking a worldwide ban on space weapons was similar to position papers issued during the Obama-Biden campaign but much less detailed and nuanced; it drew considerable attention and some criticism. By May 2009 the space part of the Defense Issues section on the White House Web site had been changed to read: "Space: The full spectrum of U.S. military capabilities depends on our space systems. To maintain our technological edge and protect assets in this domain, we will continue to invest in next-generation capabilities such as operationally responsive space and global positioning systems. We will cooperate with our allies and the private sector to identify and protect against intentional and unintentional threats to

U.S. and allied space capabilities."[1] The Obama Administration's National Space Policy (NSP), announced in June 2010, again emphasizes broad continuity between its major objectives and the overarching themes of U.S. space policy originally developed by the Eisenhower Administration, such as encouraging responsible use of space and strengthening stability in space. Other goals evolved directly from original U.S. space policy objectives, including expanding international cooperation, nurturing the U.S. space industry, and increasing assurance and resilience of mission-essential functions enabled by commercial, civil, scientific, and national security spacecraft and supporting infrastructure. The NSP indicates the United States will "ensure cost-effective survivability of space capabilities" and "develop and implement plans, procedures, techniques, and capabilities" necessary for mission assurance, including "rapid restoration of space assets and leveraging allied, foreign, and/or commercial space and nonspace capabilities to help perform the mission." There are also some areas of new or changed emphasis, such as the more enthusiastic approach toward TCBMs, including "concepts for space arms control if they are equitable, effectively verifiable, and enhance the national security of the United States and its allies," instead of the 2006 NSP language about opposing "development of new legal regimes or other restrictions that seek to prohibit or limit U.S. access to or use of space." It is important for the United States to take a measured approach toward TSBMs by carefully assessing how its new space policy can best continue recent progress in supporting effective, sustainable, and cooperative approaches to space security. In particular, it should consider how it can most effectively build on ongoing dialogue between major space actors in several venues that emphasizes incremental, pragmatic, and technical steps, moving in a bottom-up way from small measures toward larger activities. Prime examples of this approach include the December 2007 adoption of UN General Assembly Resolution 62/101, including the Inter-Agency Debris Committee (IADC) voluntary guidelines for mitigating space debris, the December 2008 Council of the European Union draft Code of Conduct for outer space activities, and the September 2010 revised draft EU Code.

Unfortunately, the new NSP falls short of appropriately and comprehensively addressing all of the NSS challenges the United States currently faces. While more stress on space cooperation is useful, the new policy overcorrects the competitive tone in the 2006 NSP by emphasizing just the cooperative dimensions of space and avoiding the reality that space is inherently a domain of both cooperation *and* competition, as states and other actors pursue their economic and security interests. A Pollyannaish approach in American declaratory policy cannot change the essential nature of human interaction in space or anywhere else and is particularly misguided following the January 2007 Chinese ASAT test that played such a large role in reawakening global concerns about space as a militarily contested and competitive domain. Another inappropriate part of the new policy states that space stability and sustainability are vital national interests. The United States has a strong interest in developing and maintaining space activities in stable and sustainable ways, but to enumerate these particular objectives as vital national interests—a term of art the United States has traditionally reserved for its most important interests to signal that it will use military force as needed in their defense— inappropriately links these broad objectives to the use of military force, implies it is within the power of the United States to maintain space stability and sustainability, and erodes the meaning of this term. Finally, the NSP fails to address how the United States will improve top-level management and organizational structures, provide clear lines of authority and responsibility, or ensure they have the durability needed to affect change, despite the fact that structural deficiencies have been a consistent theme of almost every commission studying NSS issues, and candidate Obama's pledge to reestablish a space council at the White House.

Efforts to craft comprehensive, formal, top-down space arms control or regulation continue to face the same significant problems that have overwhelmed attempts to develop such mechanisms in the past. The most serious of these problems include: disagreements over the proper forum, scope, and object for negotiations; basic definitional issues about what is a "space weapon" and how they might be categorized as offensive or defensive and stabilizing or destabilizing; and daunting concerns about whether adequate monitoring and verification mechanisms can be found for any comprehensive and formalized TCBMs. These problems relate to a number of very thorny, specific issues such as whether the negotiations should be primarily among only major spacefaring actors or more multilateral, what satellites and other terrestrial systems should be covered, and whether the object should be control of space weapons or TCBMs for space; the types of TCBMs that might be most useful (e.g., rules of the road or keep out zones) and how these approaches might be reconciled with the existing space-law regime; and verification problems such as how to address the latent or residual ASAT capabilities possessed by many dual-use and military systems or how to deal with the significant military potential of even a small number of covert ASAT systems.

New space system technologies, continuing growth of the commercial space sector, and new verification and monitoring methods interact with these existing problems in complex ways. Some of the changes would seem to favor TCBMs, such as better radars and optical systems for improved SSA, attribution, and verification capabilities; technologies for better space system diagnostics; and the stabilizing potential of redundant and distributed space architectures that create many nodes by employing larger numbers of smaller and less expensive satellites. Many other trends, however, would seem to make space arms control and regulation even more difficult. For example, micro- or nano-satellites might be used as virtually undetectable active ASATs or passive space mines; proliferation of space technology has radically increased the number of significant space actors to include a number of non-state actors that have developed or are developing sophisticated dual use technologies such as autonomous rendezvous and docking capabilities; satellite communications technology can easily be used to jam rather than communicate; and growth in the commercial space sector raises issues such as how quasi-military systems could be protected or negated and the unclear security implications of global markets for dual-use space capabilities and products.

There is disagreement about the relative utility of top-down versus bottom-up approaches to developing space TCBMs and formal arms control but, following creation of the OST regime, the United States and many other major spacefaring actors have tended to favor bottom-up approaches, a point strongly emphasized by U.S. Ambassador Donald Mahley in February 2008: "Since the 1970s, five consecutive U.S. administrations have concluded it is impossible to achieve an effectively verifiable and militarily meaningful space arms control agreement" (Mahley 2008). Yet this assessment may be somewhat myopic since strategists need to consider not only the well-known difficulties with top-down approaches but also the potential opportunity costs of inaction and recognize when they may need to trade some loss of sovereignty and flexibility for stability and restraints on others. Since the United States has not tested a kinetic energy ASAT since

September 1985 and has no program to develop such capabilities, would it have been better to foreclose this option in order to purse a global ban on testing kinetic energy ASATs, and would such an effort have produced a restraining effect on Chinese development and testing of ASAT capabilities? This may have been a lost opportunity to pursue legal approaches but is a complex, multidimensional, and interdependent issue shaped by a variety of other factors such as inabilities to distinguish between ballistic missile defense and ASAT technologies, reluctance to limit technical options after the end of the Cold War, emergence of new and less easily deterred threats, and the demise of the Anti-Ballistic Missile (ABM) Treaty.

Global licensing and export controls for space technology have often been developed and implemented in inconsistent and counterproductive ways. It is understandable that many states view space technology as a key strategic resource and are very concerned about developing, protecting, and preventing the proliferation of this technology, but the international community, and the United States in particular, needs to find better legal mechanisms to balance and advance objectives in this area. Many current problems with U.S. export controls began after Hughes and Loral worked with insurance companies to analyze Chinese launch failures in January 1995 and February 1996. A congressional review completed in 1998 (Cox Report) determined these analyses violated the International Traffic in Arms Regulations (ITAR) by communicating technical information to the Chinese. The 1999 National Defense Authorization Act transferred export controls for all satellites and related items from the Commerce Department to the Munitions List administered by the State Department. The stringent Munitions List controls contributed to a severe downturn in U.S. satellite exports. To avoid these restrictions, foreign satellite manufacturers, beginning in 2002 with Alcatel Space (now Thales) and followed by European Aeronautic Defense and Space (EADS), Surrey Satellite Company, and others replaced all U.S.-built components on their satellites to make them "ITAR-free."

There are two key reasons why the United States should move away from the priorities in its current space export control regime: first, an overly broad approach that tries to guard too many things dilutes monitoring resources and actually results in less protection for "crown jewels" than does a focused approach; and second, a more open approach is more likely to foster innovation, spur development of sectors of comparative advantage, and improve efficiency and overall economic growth. Congress and the Obama Administration have begun to reevaluate current U.S. export controls and should adjust laws and policies accordingly. Excellent starting points are the recommendations for rebalancing overall U.S. export control priorities in the congressionally mandated National Academies of Science 2009 study. In addition, the United States should implement key recommendations from the Center for Strategic and International Studies 2008 study on the space industrial base such as removing from the Munitions List commercial communications satellite systems, dedicated subsystems, and components specifically designed for commercial use.

Space and Missile Defense

The end of the Cold War changed but did not eliminate desires for ballistic missile defenses (BMD). As technology improves and proliferates, there is growing worldwide potential for "high ground" space systems such as space-based ASATs and space-based ballistic missile defenses. These developments also place increasing strains on the mutual assured destruction (MAD) construct underlying the Anti-Ballistic Missile (ABM) Treaty. Three developments during the Clinton Administration were most important in shaping the policy environment for decisions on space-based defenses; the first two moved the United States away from deploying BMD while the last moved deployment closer. Early in his first term, President Clinton reflected traditional Democratic Party ambivalence toward strategic defenses by transforming the Strategic Defense Initiative Organization (SDIO) into the Ballistic Missile Defense Organization (BMDO) and changing the direction of many programs along with this semantic change. Most importantly, BMDO moved away from the priorities of SDIO and placed its major focus on developing theater missile defenses (TMD) rather than on national missile defenses (NMD). In concert with this reordering of priorities, BMDO ended SDIO programs to develop and deploy the Brilliant Pebble (BP) space-based kinetic kill vehicle as the mainstay of the Global Protection Against Limited Strikes (GPALS) architecture, deemphasized sea-based systems built around Aegis cruisers and destroyers, and ended negotiations on managing a cooperative transition to defense deployments with the Russians. These changed priorities moved

BMDO almost exclusively into developing land-based kinetic-kill terminal defense systems—an area that may be the most politically acceptable and easiest from an arms-control perspective but is arguably the least effective and most technologically challenging for defense systems.

The Clinton Administration also changed the political and legal context for defensive systems in several ways as it was reorienting the technical focus of the program. First, following the breakup of the Soviet Union, it used the Treaty's Standing Consultative Commission (SCC) to negotiate a multilateralization to include Russia, Ukraine, Belarus, and Kazakhstan as states parties to the treaty. Second, as announced at the March 1997 Helsinki Summit, the Clinton Administration negotiated a demarcation agreement intended to strengthen and maintain the ABM Treaty by distinguishing between TMD and NMD. Operationally, the demarcation meant that defensive systems with velocities of less than three kilometers per second (3 km/s) were compliant if tested against targets with velocities of less than 5 km/s and ranges under 3,500 km. In addition, the Helsinki agreement also expanded the treaty by banning all types of space-based TMD.

The third set of major developments that shaped missile defense came at the end of the Clinton Administration and included: the June 1999 Cologne Joint Statement by Presidents Clinton and Boris Yeltsin that the United States and Russia would negotiate on modifications or amendments to the ABM Treaty allowing the United States to deploy a more robust NMD system; the congressional declaration that it is the policy of the United States to deploy an NMD system "as soon as technologically possible"; and the reorientation of the Clinton Administration's so-called 3+3 program for NMD. The 3+3 program originally called for accelerated research and testing so that, if warranted by the threat and technological progress, a decision to deploy NMD could be made in June 2000 and the system deployed by 2003. Like almost everything associated with missile defense, the 3+3 program became highly politicized. Supporters of NMD criticized it for not being a development effort commensurate with the threat and because it lacked a specific commitment to deploy NMD; critics of NMD argued that the technology to support deployment was immature and opposed this approach because it undermined the ABM Treaty and START II.

George W. Bush's Administration moved rapidly away from the ABM Treaty and deployed the first U.S. operational BMD systems since 1975. During the course of three meetings with President Vladimir Putin in 2001, Bush made it clear his administration intended to move forward with defenses, preferably in cooperation with the Russians and within the confines of modifications to the ABM Treaty, but he also indicated he would act unilaterally and withdraw from the treaty if necessary. Bush made a comprehensive statement on the ABM Treaty and nuclear deterrence at National Defense University on May 1, 2001, calling for "a new framework that allows us to build missile defenses to counter the different threats of today's world. To do so we must move beyond the constraints of the 30-year-old ABM Treaty. This treaty does not recognize the present or point us to the future. It enshrines the past" (Bush 2001). He also specifically linked building defenses with offensive strategic nuclear force reductions as part of the new framework and noted some of the potential advantages of boost phase intercepts but avoided even mentioning space-based BMD systems. On December 13, 2001, the Bush Administration provided Russia formal notification of its withdrawal from the treaty; in accordance with Article XV, the effective date of withdrawal was six months later. In addition, Secretary of Defense Donald Rumsfeld on January 2, 2002, approved a major restructuring of the Ballistic Missile Defense Organization that included a name change [to the Missile Defense Agency] and created a leaner process for developing and fielding missile-defense capabilities. The Bush Administration formally announced its policy on ballistic missile defense on May 20, 2003, emphasizing that changes in the global security environment caused by the end of the Cold War and the 9/11 attacks as well as a growing number of missile and weapons of mass destruction (WMD) threats required a different approach to deterrence and new tools for defense. Among the most worrisome events in recent missile and WMD proliferation are the North Korean Taepodong-2 launch in July 2006 and nuclear-weapon test in October 2006; continuing Iranian testing of a number of increasingly longer range missiles such as the Shahab-3, satellite launch in February 2009, and unabated major effort to develop nuclear weapons; as well as cooperation between these states and other bad actors. Even more disturbing is Pyongyang's demonstrated willingness to sell every weapon it has developed to any actor able to pay.

During the Bush Administration, MDA took a three-part approach to: maintain and sustain an initial capability to defend the U.S., allies, and our deployed forces against rogue attacks; close the gaps and improve this initial capability; and develop options for the future. To maintain and sustain initial capability, MDA deployed two interceptor missile systems, the Standard Missile-3 (SM-3) aboard Aegis Cruisers and Destroyers and Ground Based Interceptors (GBI) at Fort Greeley in Alaska and Vandenberg AFB in California; current plans call for equipping 27 ships with SM-3 capabilities by 2013, working with the Japanese on the SM-3 Block IIA (a longer-range variant), and deploying 30 GBI in Alaska and California. Additional planned actions called for enhancing early warning radars in Alaska, California, and United Kingdom; fielding a Sea-Based X-Band Radar in the Pacific and a forward-transportable radar in Japan; and expanding command and control, battle management, and communications capabilities. Original improvement plans called for adding more Aegis interceptors, fielding four transportable Terminal High Altitude Area Defense (THAAD) units, introducing land and sea variants of the Multiple Kill Vehicle (MKV) program, upgrading the early warning radar in Greenland, and establishing a GBI site and corresponding radar capability in Europe. Finally, to create options for the future, MDA planned to continue development of the Space Tracking and Surveillance System (STSS); maintain two programs, the Kinetic Energy Interceptor (KEI) and the Airborne Laser (ABL), one of which was to be selected as the boost-phase missile-defense element by 2010; and develop a Space Test Bed to examine space-based options for expanding the coverage and effectiveness for future missile- defense systems.

Following capture of both houses of Congress by the Democrats in November 2006 and the election of Barack Obama in 2008, there have been reenergized political debates over the efficacy and wisdom of missile defenses. Thus far, however, many Democratic critiques of missile defenses in the 110th and 111th Congress have been somewhat restrained and centered more on technical issues such as cost and test performance rather than broader strategic issues. Among many controversial issues, several stand out including the European third site and systems designed for boost phase intercept. In the last years of the Bush Administration, Congress cut funding for the European third site and required DoD to complete comprehensive independent studies on its implications. In addition, the Russians repeatedly raised numerous high-level objections; there were political issues in Poland and the Czech Republic, the proposed host states for the 10 GBI and associated radar; and the North Atlantic Treaty Organization (NATO) sought to link the planned third site deployment with their active layered theater missile-defense concept. In September 2009, the Obama Administration canceled plans for the European site, indicating its capabilities could be provided by a phased, adaptive approach including deployment of Aegis ships in European waters and potential stationing of land-based SM-3s in Europe in the future.

Another politically charged and technologically challenging program is the Airborne Laser (ABL), a program that has been under development for more than a decade. In March 2007, the system had its first successful airborne test of the target illuminator, demonstrating an ability to precisely track an airborne target by measuring and compensating for atmospheric distortion. The complete system, including its megawatt-class laser, was successfully tested in February 2010 when it destroyed two out of three solid- and liquid-fueled threat-representative target missiles. Following reevaluations including the congressionally directed Ballistic Missile Defense Review, the Obama Administration has chosen to convert the ABL into a technology demonstration test bed and there are no current plans to field the system operationally. Other changes include ending the Kinetic Energy Interceptor (KEI) and the Multiple Kill Vehicle (MKV) programs; developing a land-based variant of the SM-3, and dropping proposals to develop a Space Test Bed. Opponents argue this test bed is unnecessary, will weaponize space, and destabilize relations with other major nuclear-weapon states such as Russia and China. Supporters believe the test bed is needed to test space-based interceptors (that could be similar to the Brilliant Pebbles concept first discussed in the late 1980s) under operational conditions; argue that such interceptors are required to build global boost-phase intercept capabilities; and point to China's January 11, 2007, ASAT test to reemphasize that threats are growing and that "space weapons" need not be stationed in space.

Controversial programmatic issues associated with space and missile defenses outlined above pale in comparison to strategic and diplomatic consternation over the proper role of space and missile defenses. Probably the broadest and most important point of contention concerns the quest to find a "sweet spot" where U.S. missile defenses are robust enough to assure allies and dissuade, deter, and defend against emerging threats without becoming powerful enough to have the potential to neutralize a significant portion of Russian and Chinese nuclear forces and thereby undermine concepts of strategic stability. These are complex and intangible values to be balanced and it is no surprise that sophisticated conceptual models capable of fully expressing this multidimensional problem have not yet emerged. Finding this balance is made more difficult by the growing number of nuclear- and missile-armed and potentially bad actors; the limits of bilateral-only arms control that does not address both offensive and defensive forces while decreasing numbers of deployed nuclear forces between the former superpowers; the growing accuracy of conventional long-range strike systems and their emerging potential in some scenarios to mount a disarming first strike against nuclear forces; and the increasing potential of defensive systems and space-based defenses in particular. Other analysts question whether it is possible or desirable to find this sweet spot because they do not wish to delay robust defenses against bad actors for a quixotic quest for balance and stability or to replicate Cold War paradigms based on mutual vulnerability with every emerging WMD actor. Given these difficulties, it seems unlikely that a balance acceptable to all major nuclear-weapon states can be found unless the number of potential rogue actors with WMD and ballistic missiles can be reduced or the perceived need for mutual vulnerability between nuclear-armed actors can be lessened.

A second set of highly contentious issues relate to the operational benefits and drawbacks of space-based, boost-phase missile defenses versus concerns over weaponization of space and creation of space debris. Space basing would provide a number of potentially significant advantages for global boost-phase defense including always-deployed global coverage that precludes the need for crisis deployments into contested regions or littorals, rapid reaction times, and equal access to all potential launch sites. Boost-phase intercept allows engagement of missile targets during their slowest, most visible, and most vulnerable phase of flight, and has the additional benefit that any WMD aboard the missile may fall back onto the territory of the attacker. Of course, there are also daunting technical and programmatic challenges associated with boost-phase defense including the requirement to engage fleeting targets very rapidly and the potential need to pre-delegate engagement authority to the defensive systems rather than being able to maintain human decision makers in the loop under certain conditions; the costs and technical capabilities of the defenses; the absentee ratio of defenses and the potential ability of attackers to saturate defenses via salvo launches or other techniques; and the costs to boost, maintain, and replenish such a system. At the strategic level, space-based defenses raise concerns about weaponizing space due to the latent ASAT capability any missile-defense system is likely to have. Satellites are generally fragile and travel along predictable orbital paths; any defensive system capable of engaging missiles in the boost, post-boost, or midcourse phases of flight will very likely have considerable latent ASAT capabilities and it is difficult to see how such capabilities could be engineered out of a defensive system. Moving toward resolution of this strategic-level concern will require balancing the costs and benefits of defenses, especially against potential bad actors or new and unexpected threats, with concerns about space weaponization and satellite vulnerabilities. As in the case of the sweet spot issue above, it is unlikely that major actors will value these concerns in the same ways or be able to resolve these issues easily. In addition, there is also the potential that testing or use of space-based defenses could create space debris although the kinematics of interception for boost-phase defenses generally does not create large amounts of orbital debris and are especially unlikely to create long-lived debris, since neither the target nor interceptor is traveling on an orbital trajectory. Engagement of satellites or otherwise testing against objects with orbital velocities may create long-lived debris and the need for such actions should be balanced against the hazards to all space actors created by the debris.

A final set of contentious issues regarding space and missile defenses concerns the definition of a space weapon and the potential utility of controlling or regulating such weapons versus the ability of a wide range of systems to produce significant weapons effects in space or against ballistic missiles, even if they are not based in space. Simply put, since so many systems in a variety of basing modes can have effects against ballistic missiles and especially space systems, it has proven very difficult to define what a space weapon is or to control such systems. The United States is developing dedicated missile-defense systems that are land-, sea-, and air-based but should any of these systems be considered space weapons? In other words, does the basing mode matter and, if so, how? These issues are unclear but the basing mode of a system does not seem to be nearly as important as the effects it can create and the ways in which it might be used. Of course, as illustrated by the January 2007 and January 2010 Chinese hit-to-kill tests, the United States is not the only actor creating such dedicated capabilities. All the residual ASAT capabilities possessed by a wide range of systems that were not designed for this purpose greatly broaden and complicate this issue. Every space object that can transmit or maneuver has some potential to interfere with, damage, or even destroy other space objects by colliding with them. Problems in controlling residual ASAT capabilities would seem to be exacerbated by a number of trends including rapid growth in the number of commercial assets in space, movement toward smaller satellites that are more difficult to find and track, and wider development of autonomous rendezvous, proximity, and docking capabilities, including by commercial actors. In addition to all these problems primarily in space there are also numerous ways to cause interference with or disrupt satellite up- and downlinks or to attack or otherwise disrupt satellite telemetry, tracking, and control facilities on the ground.

Finally, even if all these definitional issues and problems with residual and latent ASAT capabilities could be addressed with some sort of controls, it is not clear that such controls would necessarily produce greater stability. In 1986, Ashton Carter explained the paradox of ASAT arms control—the idea that because space systems cannot be divorced from the strategic balance on Earth or the natural dialectic between offensive and defensive forces, an effective ban on ASAT weapons might make space safer for development of major space-to-Earth strike systems and result in less strategic stability. More recently, Michael O'Hanlon has questioned the need for space arms control and the desirability of granting sanctuary status to targeting systems that operate from space. Cumulatively, these factors indicate that movement toward effective and stabilizing control of space weapons and missile defenses will remain a daunting challenge and that it is hubris to suggest otherwise.

References

Biddle, S. 2004. *Military Power: Explaining Victory and Defeat in Modern Battle*. Princeton, NJ: Princeton University Press.

Bush, G.W. May 1, 2001. Remarks by the President to Students and Faculty at National Defense University. Ft. Leslie J. McNair, Washington, D.C.

Defense Science Board/Air Force Scientific Advisory Board Joint Task Force. May 2003. Report on the Acquisition of National Security Space Programs. Washington, DC: Defense Science Board.

Lupton, D. E. 1988. *On Space Warfare: A Space Power Doctrine.* Maxwell AFB, AL: Air University Press.

Mahley, D. A. February 1, 2008. "Remarks on the State of Space Security." The State of Space Security Workshop, Space Policy Institute, George Washington University, Washington, D.C.

Office of Science and Technology Policy. May 13, 2003. "Fact Sheet: U.S. Commercial Remote Sensing Space Policy." Washington, DC: The White House.

Office of Science and Technology Policy. October 6, 2006. "U.S. National Space Policy." Washington, DC: The White House.

Note

1. These statements appeared on the Defense Agenda section of the White House Web site, www.whitehouse.gov.

3

Worldwide Perspectives

As the most important space actor, the United States and its space policies play a foundational role and have come under increased scrutiny by other space actors worldwide due to the rising importance of and reliance on space capabilities for sustainable security and generating wealth. Both developing and established space powers as well as the swelling ranks of all actors that use space capabilities are increasingly interested in how domestic and international space policies are developed and implemented. This chapter reviews issue areas that are particularly important for international space policy including space law, space weaponization, space situational awareness and debris, and the emergence of China as a major space power.

Space Law and the Advancement of Spacepower

Space law has and should continue to play an essential role in the evolution of spacepower. During the 1960s, the superpowers and other emerging spacefaring states negotiated a far-reaching and forward-thinking Outer Space Treaty (OST) that continues to be perhaps the single most important means of providing structure and predictability to humanity's interactions with the cosmos. Although there is some substance to arguments that the OST only precludes those military activities that were of little interest to the superpowers and does not bring much clarity or direction to many of the most important potential space activities, the treaty nonetheless provides a solid and comprehensive foundation upon which to build additional legal structures needed to advance space-power. Spacefaring actors can most effectively improve on this foundation through a number of actions including further developing and refining the OST regime, adapting the most useful parts of analogous regimes such as the Law of the Sea and Seabed Authority mechanisms, and rejecting standards that stifle innovation, inadequately address threats to humanity's survival, or do not provide opportunities for rewards commensurate with risks undertaken. The three sections below explore other specific ways improvements in space law may contribute to: furthering the quest for sustainable space security, enabling more direct creation of wealth in and from space, and ultimately improving the odds for humanity's survival by helping to protect the Earth and space environments.

While desires for better refined space law to advance space-power may be clear, progress toward developing and implementing improvements is not likely to be fast or easy. Terrestrial law evolved fairly steadily and has operated over millennia. Space law, by contrast, is a relatively novel concept that rapidly emerged within a few years of the opening of the space age and thereafter greatly slowed. It is likely there will be growing pressure for space law to provide greater predictability and structure despite the fact that it can be very difficult to establish foundational legal elements for the cosmic realm such as evidence, causality, attribution, and precedence as well as discouraging attributes associated with spacepower that include very long timelines and prospects for only potential or intangible benefits.

It is prudent for spacefaring actors to transcend traditional categories and approaches by considering sustainable security in novel, broad, and multidimensional ways. In addition, it is often not practical or even possible to examine space law in discrete ways by delineating between legal, technical, and policy considerations or between terrestrial and space security concerns. Small, incremental steps are the most pragmatic way to develop and implement more effective space law and the process should first focus on improving and refining the foundation provided by OST regime.

Most spacefaring actors understand the merits and overall value of the OST regime; they are much more interested in building upon this foundation than in creating a new structure. As the most important first steps toward further developing space law, the international community should consider better ways to achieve more universal adherence to the regime's foundational norms and embed all important spacefaring actors more completely within the regime. Beginning work to include major non-state actors in more explicit ways could prove to be a difficult undertaking that would require substantial expansion of the regime and probably should be approached incrementally. Fortunately, the security dimensions of the regime have opened windows of opportunity, and important precedents have been set by expanding participation in the United Nations Committee on the Peaceful Uses of Outer Space (UNCOPUOS) and the World Radio Conferences of the International Telecommunications Union (ITU) to include non-state actors as observers or associate members. Some form of two-tiered participation structure within the OST regime might be appropriate for a number of years, and it may prove impractical to include non-state actors in a formal treaty, but steps toward expanded participation should begin now, both to capture the growing spacepower of non-state actors and to harness their energy in helping achieve more universal adherence to the regime. Perhaps most importantly, these initial steps should help promote a sense of stewardship for space among more actors and increase attention on those parties that fail to join or comply with these norms. Other specific areas within the OST regime that should be better developed, perhaps through creation of a standing body with implementation responsibilities, include the Article VI obligations for signatories to authorize and exercise continuing supervision over space activities and the Article IX responsibilities for signatories to undertake or request appropriate international consultations before proceeding with any activity or experiment that would cause potentially harmful interference.

In seeking better ways to harvest energy and create wealth in and from space, spacefaring actors should consider steps such as reducing liability concerns and clarifying legal issues. Of course, as with security, a range of objectives and values are in tension and require considerable effort to change or keep properly balanced. The OST has been extremely successful thus far with respect to its primary objective of precluding replication of the colonial exploitation that plagued much of Earth's history. The international community should now consider whether the dangers posed by potential cosmic land grabs continue to warrant OST interpretations that may be stifling development of spacepower, and, if these values are found to have become imbalanced, how impediments might best be reduced. Spacefaring actors should again use an expansive approach to consider how perceived OST restrictions and the commercial space sector have evolved and might be further advanced in a variety of ways.

While the OST has thus far been unambiguous and successful in foreclosing sovereignty claims and the ills of colonization, it has been less clear and effective with respect to de facto property rights and other liability and commercialization issues. OST language, negotiating history, and subsequent practice do not preclude some level of commercial activity in space and on celestial bodies, but various articles of the OST support different interpretations about the potential scope and limitations on this activity. The treaty most clearly allows those commercial activities that would be performed to support exploration or scientific efforts. It is far more problematic with respect to commercial space activity that would result in private gain or not somehow equitably distribute gains among all states. Even if it were found that commercial activities would not "appropriate" space resources, however that might be defined, it would be difficult to reconcile such activity with the spirit of the OST regime, especially since the regime provides no guidance on how private or unequal gains might be distributed.

In addition to clarifying potential property rights and wealth distribution mechanisms, consideration should be given to reevaluating liability standards. The OST and 1972 Liability Convention establish two distinct liability structures: launching states are absolutely liable to pay compensation for any damages caused by space objects on Earth or to aircraft in flight but are only liable for damages caused in space by space objects if found to be negligent. A challenge for the international community is how best to evolve the existing space-law regime based on either absolute liability or fault/negligence, depending upon the location of the incident, into a structure that might provide enough clarity to help establish liability for damages in space and perhaps provide better incentives for commercial development.

Additional interpretation issues stem from the fact that OST is embedded within a larger body of international law and that broad regime is evolving, sometimes in ambiguous and contradictory ways. Elements within this large regime are of unclear and unequal weight: the Moon Agreement with its Common Heritage of Mankind (CHM) approach to communal property rights and equally shared rewards undoubtedly has some effect in advancing the CHM principle in both formal and customary international law. At the level of formal international law, however, the Moon Treaty falls well short of the OST due to its lack of parties, especially among major spacefaring states, especially in contrast to the OST, a treaty that has been ratified by some 94 states and in force for over 40 years.

Most fundamentally, however, the current lack of clarity within space law about property rights and commercial interests is the result of both space law and space technology being underdeveloped and immature. Of course, there is also a "chicken-and-egg" factor at work since actors are discouraged from undertaking the test cases needed to develop and mature the regime because of the immaturity of the regime and their unwillingness to develop and employ improved technologies and processes as guinea pigs in whatever legal processes would be used to resolve property rights and reward structures. The most effective way to move past this significant hurdle would be to create more clear mechanisms for establishing property rights and processes by which all actors, especially commercial actors, could receive rewards commensurate with the risks they undertake. In addition, any comprehensive reevaluation of space property rights and liability concerns should also consider how these factors are addressed in analogous regimes such as the Seabed Authority in the Law of the Sea Treaty. Unfortunately, however, there are also several problems with attempting to draw from these precedents. First, several of the analogous regimes like the Law of the Sea build from CMH premises in several ways and it is not clear this approach is entirely applicable or helpful when attempting to sort through how the OST should apply to issues like property rights and reward structures. Second, while these analogous regimes are undoubtedly better developed than the OST and have a significant potential role in providing precedents, today they are still somewhat underdeveloped and immature with respect to their application in difficult areas such as property rights and reward structures, again limiting the current utility of attempting to draw from these precedents.

Provisions of the OST regime are probably the most important factors in shaping commercial space activity, but they are clearly not the only noteworthy legal and policy factors at work influencing developments within this sector. Legacy legal and policy structures developed during the Cold War were probably adequate for the amount of commercial space activity during that period, but it is far from clear they will be sufficient to address the significant and sustained increase in commercial space activity since that time. It would be helpful if governments, and the U.S. government in particular, could more explicitly develop and consistently implement legal structures and long-term policies that would better define and delineate between those space activities that ought to be pursued by the private and public sectors as well as more intentionally and consistently develop the desired degree of international cooperation in pursuing these objectives.

The United States and other major spacefaring actors lack, but undoubtedly need, much more open and comprehensive visions for how to develop spacepower. The process for developing spacepower needs to be emphasized, become more intentional and formalized, and be supported by an enduring organizational structure that includes the most important stakeholders in the future of spacepower. Legal structures should be a foundational part of creating and implementing the vision to develop spacepower, but the approach should be broader, focused on opening space as a medium for the full spectrum of human activity and commercial enterprise, and all actions governments can take to promote and enable spacepower, including government development of foundational, noneconomically competitive infrastructure and technology; encouraging incentive structures (prizes, anchor-customer contracts, and favorable property or exclusion rights); and endorsing liberal regulatory regimes (port authorities, spacecraft licensing, and public-private partnerships). In addition, consideration should be given to using other innovative mechanisms and nontraditional routes to space development, including a much wider range of federal government organizations and the growing number of local or state spaceport authorities and other organizations developing needed infrastructure. Moreover, the United States should consider making comprehensive and careful exploration of the potential of space-based solar power as its leading pathfinder in creating a vision for developing spacepower. Working toward harvesting this unlimited power source in economically viable ways will require development of appropriate supporting legal structures, particularly with respect to indemnification and potential public-private partnerships.

Finally, the international community should consider making more effort toward developing space law to improve environmental sustainability and humanity's odds for survival. Space provides a unique location to monitor and potentially remediate Earth's climate. It is the only location from which simultaneous in situ observations of Earth's climate activity can be conducted and such observations are essential to developing a long-term understanding of potential changes in our biosphere. Because so much is riding on our understanding of the global climate and our potential responses to perceived changes, it is particularly important to apply apolitical standards in getting the science right and controlling for known space effects such as solar cycles when making these observations. If fears about global warming are correct and the global community wishes to take active measures to remediate these effects, space also provides a unique location to operate remediation options such as orbital solar shades.

It is also imperative that the United States and all spacefaring actors think more creatively about using spacepower to transcend traditional and emerging threats to our survival. Humanity needs more consideration of genuinely cooperative approaches for protecting the Earth and space environments from cataclysmic events such as large objects that may collide with Earth or gamma ray bursts that may have the potential to render huge swaths of space uninhabitable. Better knowledge about known threats such as near-Earth objects (NEOs) is being developed but more urgency is required. All predicted near approaches and possible NEO impacts such as the asteroid Apophis on April 13, 2029, ought to be seen as opportunities since they provide critical real-world tests for our ability to be proactive in developing effective precision tracking and NEO mitigation capabilities. In the near term, it is most important for national and international organizations to be specifically charged with and resourced to develop better understanding of NEO threats and mitigation techniques that can be effectively applied against likely impacts. Ultimately, however, we cannot know of or effectively plan for all potential threats to Earth but should pursue a multidimensional approach to develop capabilities to improve our odds for survival and one day perhaps become a multiplanetary species. The inexorable laws of physics and of human interaction indicate that we will create the best opportunities for success in improving space law by beginning long-term, patient work now rather than crash programs later. This long-term, patient approach will allow the best prospects for space law to provide a solid foundation for the peaceful advancement of spacepower.

Space Weaponization

At a fundamental level, virtually all issues of space strategy turn on broad questions related to weaponizing space such as whether space will be weaponized, how and when that might happen, which states and other actors might be most interested in leading or opposing weaponization, and how space weaponization issues might best be controlled. At the political level, there is, of course, a broad spectrum of opinion on these issues but most of the major tenets in mainstream views on weaponizing space can usefully be grouped into four major camps: Space Hawks, Inevitable Weaponizers, Militarization Realists, and Space Doves. Each of these camps is described below and they are used to analyze sources of support or opposition for attempts to control space weaponization.

Space Hawks believe that space already is or holds the potential to become the dominant source of military power. Accordingly, they advocate that the United States move quickly and directly to develop and deploy space weapons in order to control and project power from this dominant theater of combat operations. According to former Senator Bob Smith, for example, the concerted development of American space weapons "will buy generations of security that all the ships, tanks, and airplanes in the world will not provide. . . . Without it, we will become vulnerable beyond our worst fears" (Smith 1999, 33). In addition, Space Hawks often point to space-based BMD as a potentially decisive weapon capable of fundamentally reordering the strategic balance. Space Hawks tend to oppose virtually all space-related arms control or regulation because of its potential to slow or derail rapid and direct space weaponization by the United States.

Inevitable Weaponizers think that space, like all other environments humans have encountered, will eventually be weaponized. They differ from Space Hawks in two important ways: they are not convinced that space weaponization would be beneficial for U.S. or global security and they are unsure that space will prove to be the decisive theater of combat operations. The Space Commission report presents this position: "We know from history that every medium—air, land and sea—has seen conflict. Reality indicates that space will be no different. Given this virtual certainty, the U.S. must develop the means both to deter and to defend against hostile acts in and from space" (Commission to Assess National Security Space Management and Organization 2001, x). Inevitable Weaponizers take a nuanced view of space arms control and regulation. They generally support TCBMs and other mechanisms designed to slow military competition and channel it in predictable ways. But they are less supportive of broad efforts to ban space weapons because they see them as futile or even dangerous due to their potential to lull the United States into complacency or otherwise cause it to be outmaneuvered by states that successfully circumvent space weaponization accords.

Militarization Realists oppose space weaponization because they believe U.S. security interests are best served by the status quo in space. They believe that the United States has little to gain but much to lose by weaponizing space because it is both the leading user of space and, enabled by this space use, the dominant terrestrial military power. Militarization Realists also believe that if the United States takes the lead in weaponizing space, it would become easier for other states to follow due to lower political and technological barriers. They generally believe that fighting *into* space looks feasible and we should plan for the eventuality; fighting *in* space currently has little utility; while fighting *from* space looks impractical for the foreseeable future, with or without arms control. Militarization Realists may support space-related arms control and regulation that precludes other states from weaponizing or even militarizing space. Most of them believe, however, that this support must be balanced against the increased attention that formalized arms-control efforts could draw to the United States' already formidable space-enabled force enhancement capabilities and the political, military, and arms-control fallout this increased scrutiny might cause.

Finally, a wide range of organizations and viewpoints can be grouped together as Space Doves because they all oppose space weaponization for a variety of reasons including moral, arms control, conflict resolution, stability, and ideology arguments. Many Space Doves also oppose any militarization of space beyond the limited missions they see as stabilizing—national technical means (NTM), early warning, and hotline communications—because they see any military missions beyond these as the "slippery slope" to space weaponization. Space Doves emphasize that unlike the strategy for nuclear weapons, there exists no obvious strategy for employing space weapons that will enhance global stability and conform to the long-standing precedent of U.S. foreign policy attempting to avoid destabilizing situations. They also highlight the deep roots of President Eisenhower's "peaceful purposes" policy and argue that, even in the post–Cold War and post-9/11 era, there is no rationale for space weaponization strong enough to overturn the basic strategic logic America developed at the opening of the space age. Space doves support space arms control and regulation more strongly than any other camp. Since they do not believe the United States (or other states) would reap strategic benefits from weaponizing space, they are not overly concerned about the numerous arms-control challenges identified by the other camps.

Threats due to potential high-altitude nuclear detonations (HAND) are sufficiently different and potentially damaging that they warrant discussion and analysis separate from broad space weaponization issues. Just one such detonation holds the potential to disable *all* nonhardened Low-Earth Orbit (LEO) satellites. Today, it would require hundreds of billions of dollars to replace these assets; their value and the importance of the capabilities they provide are likely to continue growing; and this threat poses daunting detection, deterrence, and defense challenges—not least is the fact that such an attack would take place outside the sovereign jurisdiction of any state and perhaps not directly kill a single person. HAND and electromagnetic pulse (EMP) attacks against space and terrestrial targets such as power grids present unique asymmetric threats that may be the most potentially disruptive and dangerous ways nefarious actors could use single or small numbers of nuclear weapons. HAND promptly burnout the electronics for satellites relatively close and within line of sight of the explosion; then, in weeks to months, nonhardened LEO satellites fail due to their inability to operate in the pumped up radiation belts caused by the detonation. According to a 2001 study conducted for the Defense Threat Reduction Agency, LEO satellites can be hardened against total radiation dose failures for only 2–3 percent more than the costs they already incur to harden against natural radiation.

Congested Space: Spectrum Crowding, SSA, and Orbital Debris

The cumulative effects of greater use of space are creating an increasingly congested environment that may require increased or enhanced regulatory and remediation mechanisms. Growth in commercial space activity has exacerbated crowding of the radio spectrum for space applications and there are significant pressures on portions of the spectrum now allocated to military uses. In particular, consideration has been given to moving DoD out of the 1755 to 1850 megahertz (MHz) band and moving a large number of U.S. government users out of the 1710–1755 MHz band (an action estimated to cost more than a billion dollars) in order to auction it off for third generation (3G) telecommunications applications. It is not clear, however, that U.S. national security or even economic interests would necessarily benefit from this action; a number of studies, including an August 2001 Government Accountability Office (GAO) report, indicate that more assessment is required and call for spectrum issues to be carefully evaluated in light of emerging telecommunication technologies, reduced bandwidth requirements due to improving technology, and unclear commercial and consumer demands. Improving technologies and different orbits have lessened the effects of increased use and reduced GEO overcrowding concerns; modern GEO systems, for example, have two or fewer degrees of spacing versus three or more degrees in the past. Current trends for the space radio spectrum do not augur major changes in the regulatory structure. Moving the ITU to auctions for its coordination/registration process would undoubtedly produce greater efficiency and generate income, but these benefits would need to be balanced against the equal access concerns of the developing world and the current lack of support for moving in this direction.

One of the best ways for the United States to demonstrate leadership in fostering cooperative spacepower and help better define OST regime implementation obligations would be to share space situational awareness (SSA) data globally in more effective ways through its SSA Sharing Program or some other approach. The Secretary of Defense through U.S. Strategic Command's Joint Space Operations Center (JSpOC) at Vandenberg AFB in California is responsible the SSA Sharing Program. Following the February 2009 collision between Iridium and Cosmos satellites, there is more worldwide attention focused on space debris and spaceflight safety as well as considerable motivation for the United States to improve the program by providing SSA data to more users in more timely and consistent ways.

SSA issues are framed by specialized concepts and jargon. Conjunctions are close approaches, or potential collisions, between objects in orbit. Propagators are complex modeling tools used to predict the future location of orbital objects. Satellite operators currently use a number of different propagators and have different standards for evaluating and potentially maneuvering away from conjunctions. Maneuvering requires fuel and shortens the operational life of satellites. Orbital paths are described by a set of variables known as ephemeris data; two-line element sets (TLEs) are the most commonly used ephemeris data. Data about tracked orbital objects is usually presented in the form of a satellite catalog. The United States maintains a public catalog at www.space-track.org. Other entities maintain their own catalogs. Orbital paths constantly change, or are perturbed, by a number a factors including Earth's inconsistent gravity gradient, solar activity, and the gravitational pull of other orbital objects. Perturbations cause propagation of orbital paths to become increasingly inaccurate over time; beyond approximately four days into the future, predictions about the location of orbital objects can be significantly inaccurate.

A most useful specific goal for the SSA Sharing Program would be creation of a U.S. government operated data center for ephemeris, propagation data, and pre-maneuver notifications for all active satellites; consideration should also be given to the utility and modalities of creating or transitioning such a data center to international auspices. Users would voluntarily contribute data to the center, perhaps through a GPS transponder on each satellite, and the data would be constantly updated, freely available, and readily accessible so that it could be used by satellite operators to plan for and avoid conjunctions. In December 2009, satellite owners and operators created the Space Data Association to share SSA data. Governments face difficult legal, technical, and policy issues that inhibit progress including bureaucratic inertia; liability and proprietary concerns; nonuniform data formatting standards and incompatibility between propagators and other cataloguing tools; and security concerns over exclusion of certain satellites from any public data. Some of these concerns could be addressed by working toward better cradle-to-grave tracking of all catalogued objects to help establish the launching state and liability; using opaque processes to exclude proprietary information from public databases to the maximum extent feasible; and indemnifying program operators, even if they provide faulty data that results in a collision, so long as they operate in good faith, exercise reasonable care, and follow established procedures.

Minimizing, avoiding, and potentially mitigating orbital debris may represent the best window of opportunity for cooperative space TCBMs for the United States and the global spacefaring community. All human space activity produces orbital objects; when these objects no longer serve a useful function, they are classified as space debris. Over time, human activity has generated an increasing amount of debris from a variety of causes; the number of catalogued debris objects has gone from about 8,000 to over 21,500 during the past 20 years. Intentional and unintentional satellite explosions, spent rocket bodies, and many types of space operations and maneuvers cause space debris, but the most disturbing source of debris is deliberate hypervelocity impacts between large objects at high orbital altitudes such as the Chinese direct ascent kinetic energy ASAT weapon test on January 11, 2007. This dangerously irresponsible test created more than 3,000 tracked objects and now accounts for more than 22 percent of all catalogued objects in LEO. Unfortunately, less than 3 percent of the debris from this test has reentered the atmosphere so far, and it is estimated that many pieces will remain in orbit for decades and some for more than a century (*NASA Orbital Debris Quarterly News*, October 2010, 3). By contrast, destruction of the inoperative USA-193 satellite on February 21, 2008, occurred at a much lower altitude and did not produce long-lived debris; the last piece of catalogued debris from this intercept reentered Earth's atmosphere and burned up on October 9, 2008.

If current trends continue, there is growing risk that debris will make space, and LEO in particular, increasingly unusable. Fortunately, there is also growing awareness and earnestness across the international community in addressing this threat. Since the 1980s, the United States has led the world in publicizing risks due to orbital debris and it has made programs to mitigate debris an increasingly important part of its overall space policy. Overarching goals for spacefaring actors with respect to space debris include minimizing its creation while mitigating and remediating its effects. Key approaches to minimizing creation of debris are commercial best practices and evolving regimes such as the Inter-Agency Space Debris Coordination Committee (IADC) voluntary guidelines adopted by the UN General Assembly Resolution 62/101 in December 2007. Spacefaring actors also need to consider mechanisms to transition these voluntary guidelines into more binding standards and ways to impose specific costs such as sanctions or fines on actors that negligently or deliberately create long-lived debris. Fines could be applied toward efforts to further develop and educate spacefaring actors about the debris mitigation regime as well as to create and implement remediation techniques. An additional potential source of funding for mitigation

and remediation would be establishing auctions for the radio frequency spectrum controlled by the ITU that would be analogous
to the spectrum auctions conducted at the national level by organizations like the FCC. Finally, it must be emphasized that techniques for remediating debris using lasers or other methods are
likely to have significant potential as ASAT weapons, and very
careful international consideration should be given to how and by
whom such systems are operated.

In addition the international community should consider the
need for and implications of space traffic management systems
(STMs) that could be analogous to current air-traffic control systems. The idea for such a system is obviously related to SSA and
orbital debris, but it goes beyond to include a wide range of factors
such as: how space traffic might coordinate and be approved for
specific orbital positions, how space traffic would be located and
tracked, sanctions and liability for noncompliance and collisions
under an STM system, and how such a regime might be established and funded. As with many space-related issues, the technology to at least begin implementing such a system appears to
be closer at hand than is the political will to begin down this path.
The international community should consider very carefully how
its objectives in space might benefit or be harmed via the creation
and operation of an STM system. It is not obvious that an air-traffic-
control model is the appropriate regime for space, or that the
political and financial costs of creating and operating such a system (many of which would likely be borne by the United States)
would be outweighed by its benefits. Most of the benefits would
seem to be in the commercial and civil space sectors while the
potential drawbacks might be most pronounced for the military
and intelligence sectors. The United States most likely would not,
for example, want the ephemeris on its military and intelligence-
gathering satellites to be pre-approved and available worldwide
through an STM system. At the very least, since STMs could be
such a powerful tool for denial, deception, and even targeting, the
United States must think through very carefully exactly what type
of management regime would be most appropriate for space and
how such a regime would operate in practice.

China as an Emerging Space Power

China presents the global community with both the greatest opportunities and the most difficult challenges for space cooperation.

In October 2003, China independently launched and recovered its first taikonaut, or astronaut, becoming just the third member of an elite spacefaring club with Russia and the United States. Then in January 2007 China first successfully tested a kinetic-energy, direct-ascent ASAT weapon and again joined Russia and the United States as one of only three states known to have demonstrated this capability. China's growing power and space emphasis may become manifest in mostly peaceful and cooperative ways or may lead to increasing competition and perhaps even conflict. If the United States and others can successfully engage China in effective space cooperation, this may reduce risks of increasing competition, but all parties must avoid the mistake of treating China like the Soviet Union or seeing this relationship through the lens of the Cold War. The Soviet Union was only a military superpower, whereas China is a major U.S. trading partner and an economic superpower that recently passed Germany and Japan to become the world's second largest economy and is on a path to become larger than the U.S. economy, perhaps within only 10 years. Because of its economic muscle, China can afford to devote commensurately more resources to its military capabilities and will play a more significant role in shaping the global economic system. For example, China holds more than $1.4 trillion in foreign assets (mainly U.S. treasury notes), an amount that gives it great leverage in the structure of the system.

Like many other major spacefaring states around the world, China does not make clear distinctions between its civil and military space activities, pursuing instead many advanced and dual-use capabilities with military applications, sometimes even with foreign partners such as on the China-Brazil Earth Resources Satellite (CBERS) program. China has moved more quickly in developing a wider range of military space capabilities than any previous spacefaring state and today has deployed comprehensive space systems that are less capable but parallel those of the United States in all mission areas except for space-based missile launch detection.

China's civil space effort began in earnest in the post–Cold War era; it pursues human spaceflight and exploration for prestige and to set China apart as a great power. From the beginning, however, all Chinese space activity, including its civil space activity, has been either directly or indirectly controlled by the People's Liberation Army (PLA). Although some Chinese civil space efforts began in the 1950s and the China National Space Agency (CNSA) was established in 1993, ostensibly to direct China's civil space program, under the current bureaucratic structure and for most of its existence CNSA was a small organization with only about 250 people embedded within the Commission for Science, Technology, and Industry for National Defense (COSTIND), a higher ministerial entity that oversaw many of China's defense industries. Moreover, CNSA appears to have little decision-making authority; its main function seems to be to interface with foreign space agencies, a role similar to that played by the Ministry of Defense and other organizations within the Chinese government that present this type of facade as the way the outside world is to interact with the Middle Kingdom but can cause problems in correctly aligning counterpart organizations and decision-making structures.

Now that it has achieved its major initial prestige goals, China may become more interested in partnering on cooperative civil space efforts such as the ISS or other joint projects to pursue the ambitious exploration goals it has espoused that include orbiting a space laboratory in 2013, building a permanently inhabited space station beginning in 2016, and perhaps undertaking a lunar landing by 2020. It is not clear, however, whether China will continue to pursue civil space objectives primarily unilaterally, will work increasingly with the very diverse members of the Asia-Pacific Space Cooperation Organization (APSCO) it has established, or partner with other major space actors. If China is interested in pursuing cooperative civil space efforts with the United States, it will need to make that more clear than it did to Michael Griffin in September 2006 when he made the first visit by a NASA Administrator to China yet was granted only limited access to his counterpart space decision makers and other space personnel and facilities. The rhetoric during the October 2009 visit of the second-highest ranking PLA member, General Xu Caihou, vice chairman of the Chinese Central Military Commission, to a number of important U.S. locations, including the headquarters of Strategic Command, as well as the dialogue between Presidents Hu Jintao and Barack Obama during Obama's November 2009 visit to Beijing, and the more open visit of NASA Administrator Charles Bolden to China in October 2010, offer an opportunity to begin building cooperative space efforts and developing better space and security relationships.

Of course, it has not escaped notice worldwide that the United States has already employed and continues to develop network-enabled warfare or that space capabilities often provide the best and sometimes the only way to make these kinds of operations possible. The Chinese, in particular, have been among the most careful students of the modern system and U.S. space-enabled military operations over the last generation. They have concluded that information operations and space capabilities are required to fight and win what it refers to as "local wars under conditions of informatization" and are following their own unique path toward improved military potential while making significant efforts both to emulate and counter U.S. space capabilities.

> In the past, PLA authors acknowledged that its informa-
> tion systems were incapable of enabling it to act more
> quickly than the U.S. military and their writings focused
> more on denying space to potential adversaries. How-
> ever, as the PLA begins to contemplate using space, it rec-
> ognizes that it must not only deny the use of information
> to its opponents but also use space to facilitate its own
> operations. (Pollpeter, 2008, 26)

Leveraging its latecomer's advantage during its 10th (2001–2005) and 11th (2006–2010) Five Year Plans, China has moved more quickly in developing a wider range of space capabilities than any previous spacefaring state and today has deployed comprehensive space systems that are less capable but parallel those of the United States in all mission areas except for space-based missile launch detection. China's array of space reconnaissance systems offer increasingly precise visible, infrared, multispectral, and synthetic aperture radar imaging and include the *Ziyuan*-2 series, the *Yaogan*-1 through -11, the *Haiyang*-1 and -2 series, the CBERS series, and the *Huanjing* disaster and environmental monitoring satellite constellation. In the next decade, Beijing most likely will field radar, ocean surveillance, and high-resolution photoreconnaissance satellites. In the interim, China probably will rely on commercial satellite imagery to supplement existing coverage. For navigation and timing, the Chinese have launched four *Beidou* satellites that provide signals with 20-meter accuracy over China and surrounding areas. China also plans to complete a more advanced, accurate, and global PNT system known as *Beidou*-2 or Compass comprised of five Geostationary-Earth Orbit (GEO) satellites and 30 Medium-Earth Orbit (MEO) satellites; the Compass-M1 experimental satellite was launched in April 2007 and Compass-G1 and -G2 satellites were launched in 2009 and 2010. In addition, China has a very advanced indigenous microsatellite program with microsatellites currently deployed for technology development, imagery, remote sensing, and communications missions. Finally, China uses a wide range of foreign and domestic communications satellites, is increasing its military employment of these communications capabilities, and is moving to replace all foreign communications satellites with indigenous satellites by 2010. With launch in April 2008 of its first tracking and data relay satellite (*Tianlian* I), the Chinese have demonstrated the potential to develop a nascent real-time, global reconnaissance strike complex.

China is moving more secretly but probably even more quickly and comprehensively in developing a multidimensional program to limit or prevent the use of space-based assets by its potential adversaries during times of crisis or conflict. The PLA has deployed a variety of kinetic and nonkinetic weapons and terrestrial jammers and is also exploring other counterspace capabilities including in-space jammers, high-energy lasers, high-powered microwave weapons, particle beam weapons, and electromagnetic pulse weapons. In addition, China is researching and deploying capabilities intended to disrupt satellite operations or functionality without inflicting physical damage. The successful Chinese ASAT test of January 2007 was perhaps most notorious for its dangerous irresponsibility in creating a persistent debris cloud, but the debris the test created should not obfuscate the system's very significant strategic implications given the high-value U.S. assets it can hold at risk in LEO, difficulties in finding and tracking the road-mobile transporter-erector-launcher (TEL) for the *Dong Feng* (DF)-21 (or SC-19) intermediate-range missile that launches the ASAT, and the extremely limited protection measures the United States currently has against this capability. Moreover, the direct ascent ASAT is just one of the many types of counterspace capabilities the Chinese are developing or have already fielded; it may not even be their most threatening or pervasive capability. It is more important to consider the synergistic and tailored benefits China is likely to obtain by employing many counterspace capabilities that operate in different ways against different orbital regimes and mission areas including hundreds if not thousands of high-power mobile terrestrial jammers, high-energy lasers with precision tracking capabilities at multiple sites, and potentially sophisticated in-space jamming and negation capabilities.

Tensions between the United States and China and between Taiwan and China have been easing in a number of ways and this book is not suggesting that conflict over Taiwan is imminent.

However, many seemingly irreconcilable issues remain, including the "sacred responsibility" of the PLA in stopping independence and the "anti-succession" law passed by China's National People's Congress in March 2005; U.S. commitments under the 1979 U.S. Taiwan Relations Act to resist any force or other coercion that threatens Taiwan; and Taiwanese independence aspirations. Taiwan is still clearly the most likely flashpoint for any conflict between the United States and China. Because the PLA is continuously improving the quality and effectiveness of its overall military capabilities and China's space activities and capabilities, including ASAT programs, have significant implications for anti-access/area denial in Taiwan Strait contingencies and beyond, military analysts must continually assess the correlation of forces in this scenario and statesmen must remain aware of its implications. Michael O'Hanlon's 2004 assessment is both reassuring and sobering, especially given the continuing and accelerating progress of PLA modernization and the considerable stresses placed on U.S. forces by ongoing operations in Iraq and Afghanistan:

> It is doubtful that trends in space capabilities or other aspects of defense modernization will radically alter the military balance in the next decade or so. The size and caliber of the U.S. military is sufficient that, even if China were able to close the technological gap and have the potential to cause substantial losses to the United States in a war over Taiwan, the American armed forces would surely prevail. The United States could lose a carrier or two and still maintain overwhelming military superiority in the region. (O'Hanlon 2004, 27)

In a presentation at the Naval War College in April 2009 Secretary of Defense Robert Gates found adversary development of anti-access/area denial capabilities more troubling and noted

> a particular concern with aircraft carriers and other large, multi-billion dollar blue-water surface combatants— where the loss of even one ship would be a national catastrophe. We know other nations are working on ways to thwart the reach and striking power of the U.S. battle fleet—whether by producing stealthy submarines in quantity or developing anti-ship missiles with increasing range

and accuracy. We ignore these developments at our peril.
(Gates 2009)

A large number of factors and complex interrelationships are
involved, but all Taiwan scenarios are fundamentally shaped by
a small number of geopolitical factors including the very close
proximity of the theater of operations to China and its extreme dis-
tance from the United States, very limited basing options for U.S.
forces in this region, and the increasing vulnerability of all fixed
and even some mobile targets to attack from a growing number of
long-range precision strike forces. These factors combine to make
the effectiveness of U.S. aircraft carrier battle groups a most important variable in any Taiwan scenario. A key objective for China
is to find and strike carrier battle groups as far away from Taiwan
as possible; keeping them out of the main fight or at least primarily focused on self-defense. For the United States and Taiwan, key
objectives include finding and striking a large percentage of landing craft and transport aircraft before they can lodge and sustain
an overwhelming number of ground forces on Taiwan.

Space and counterspace capabilities play an increasingly
important role for both sides in this scenario. For China, space
forces, and space ISR in particular, are needed to find, fix, track,
target, engage, and assess strikes on carrier battle groups in near
real time. Space links considered necessary for day-night, inclem-
ent weather, and near real-time operation of this kill chain include
high-resolution imagery, tracking and data relay, synthetic aperture radar, wake tracking, and electronic intelligence—all capabilities the Chinese appear to have increasingly emphasized. It is
not yet clear that China has networked together all the capabilities required for long-range precision strikes against carrier battle
groups let alone what the effectiveness of Chinese forces so employed might be, even before they are attrited by the concentric
layers of defenses around carrier battle groups. Nonetheless, it is
apparent that Chinese capabilities for long-range precision strikes
against ships have improved significantly; U.S. forces are threat-
ened as they approach what the Chinese call the second island
chain that includes Guam, and operate at growing peril the closer
they come to Taiwan and the first island chain. The increasingly
potent antiaccess strike forces the Chinese have deployed or are
developing include large numbers of highly accurate cruise missiles, such as domestically produced ground-launched DH-10
land attack cruise missiles, SS-N-22/Sunburn and SS-N-27B/Sizzler

supersonic antiship cruise missiles mounted on *Sovremennyy*-class guided missile destroyers and Kilo-class diesel electric submarines acquired from Russia, as well as an antiship ballistic missile based on a variant of the DF-21 that has a range in excess of 1,500 kilometers and highly accurate maneuvering reentry vehicles with conventional warheads and terminal-sensitive penetrating submunitions designed to destroy enemy carrier-borne planes, the control tower and other easily damaged and vital positions. It is also a near certainty that China would mount large-scale counterspace operations, perhaps even as a precursor to other attacks, in any Taiwan scenario. Chinese counterspace operations would likely concentrate on cyber and electronic warfare attacks against U.S. communications and positioning, navigation and timing (PNT) capabilities using terrestrial, airborne, seaborne, and perhaps in-space jammers or ASAT systems. In addition, the Chinese could use their direct-ascent ASAT and high-energy lasers to attack U.S. ISR assets in LEO and it is unlikely that either preemptive or reactive maneuvering of these assets would be able to protect them or ensure they could collect on assigned targets.

U.S. and Taiwanese space capabilities and counterspace operations are also critically important, would be highly stressed in defending Taiwan, and would be tested in novel ways since the United States has not yet fought against a space-enabled, near-peer military power. All U.S. space-enabled force enhancement capabilities—ISR, missile warning and attack assessment, communications, PNT, and environmental monitoring—would be challenged in attempting just to establish and maintain a kill chain for the thousands of fixed and mobile targets in the Taiwan theater even without enemy countermeasures; in a degraded electronic warfare environment and under direct attack, their efficacy is likely to be significantly reduced. Under these conditions, projecting strike assets into the theater and maintaining an effective kill chain, especially against the many small and fleeting, mobile targets presented by Chinese landing craft and aircraft, would be a daunting challenge. The United States would also engage in counterspace operations, primarily to disrupt PNT and command and control of landing forces as well as in attempts to deny Chinese ability to track and target carrier battle groups. With respect to the latter counterspace objective in particular, it is noteworthy that O'Hanlon believes the United States could be quite hard pressed to disrupt Chinese ability to target carriers in a Taiwan scenario without ASAT capabilities such as those it demonstrated in February 2008 when an Aegis

Cruiser used a Standard Missile-3 to destroy the inoperative USA-193 satellite just prior to reentry.

Current efforts to craft comprehensive top-down space arms control or regulation with the Chinese bilaterally, or among all spacefaring actors, still face all of the significant problems that plagued attempts to develop such mechanisms in the past. The most serious of these problems include: disagreements over the proper scope and object of negotiations; basic definitional issues about what is a space system and how they might be categorized as offensive or defensive and stabilizing or destabilizing; and daunting questions concerning how any agreement might be adequately verified. These problems relate to a number of very thorny specific issues such as whether the negotiations should be bilateral or multilateral, what satellites and other terrestrial systems should be covered, and whether the object should be control of space weapons or TCBMs for space; questions concerning which types of TCBMs such as rules of the road or keep out zones, for example, might be most useful and how these might be reconciled with existing space law such as the OST; and verification problems such as how to address the latent or residual ASAT capabilities possessed by many dual-use or military systems or deal with the significant military potential of even a small number of covert ASAT systems. New space system technologies, continuing growth of the commercial space sector, and new verification technologies interact with these existing problems in complex ways. Some of the changes would seem to favor arms control and regulation, such as better radars and optical systems for improved SSA and verification capabilities, technologies for better space system diagnostics, and the stabilizing potential of microsatellite-based redundant and distributed space architectures. Many other trends, however, would seem to make space arms control and regulation even more difficult. For example, micro- or nanosatellites might be used as virtually undetectable active ASATs or passive space mines; proliferation of space technology has radically increased the number of significant space actors to include a number of non-state actors that have developed or are developing sophisticated dual use technologies such as autonomous rendezvous, proximity, and docking capabilities; and growth in the commercial space sector raises issues such as how quasi-military systems could be protected or negated and the unclear security implications of global markets for dual-use space capabilities and products.

Despite these difficulties, the Chinese apparently disagree with pursuing only bottom-up approaches and, in ways that seem both shrewd and hypocritical, are currently developing significant counterspace capabilities while simultaneously advancing various top-down proposals in support of prevention of an arms race in outer space (PAROS) initiatives and moving ahead with the joint Chinese-Russian draft treaty on Prevention of Placement of Weapons in Outer Space (PPWT) introduced at the Conference on Disarmament in February 2008. If the Chinese are attempting to pursue a two-track approach to space arms control, they need to present that argument to the international community much more explicitly. The current draft PPWT goes to considerable lengths in attempting to define space, space objects, weapons in space, placement in space, and the use or threat of force, but there are still very considerable definitional issues with respect to how specific capabilities would be classified. An even more significant problem relates to all the terrestrial capabilities that are able to eliminate, damage, or disrupt the normal function of objects in outer space such as the Chinese direct ascent ASAT. One must question the utility of a proposed agreement that does not address the significant security implications of current space system support for network enabled terrestrial warfare, does not deal with dual-use space capabilities, seems to be focused on a class of weapons that does not exist or at least is not deployed in space, is silent about all the terrestrial capabilities that are able to produce weapons effects in space, and would not even ban development and testing of space weapons, only their use. Given these weaknesses in the PPWT, it seems plausible that it is designed as much to continue political pressure on the United States and derail U.S. missile-defense efforts as it is to promote sustainable space security.

Cooperative space ventures or TCBMs that have been proposed and are worthy of further consideration include: inviting a taikonaut to fly on one of the remaining space shuttle missions and making repeated, specific, and public invitations for the Chinese to join the ISS program and other major cooperative international space efforts. The world spacefaring community could also work toward developing nonoffensive defenses of the type advocated by Philip Baines. Kevin Pollpeter explains how China and the United States could cooperate in promoting the safety of human spaceflight and "coordinate space science missions to derive scientific benefits and to share costs. Coordinating space science missions with separately developed, but complementary space assets, removes the chance of sensitive technology transfer and allows the two countries to combine their resources to achieve the same effects as jointly developed missions" (Pollpeter 2008,

48–50). Michael Pillsbury outlined six other areas where U.S. experts could profitably exchange views with Chinese specialists in a dialogue about space weapons issues: "reducing Chinese misperceptions of U.S. Space Policy, increasing Chinese transparency on space weapons, probing Chinese interest in verifiable agreements, multilateral versus bilateral approaches, economic consequences of use of space weapons, and reconsideration of U.S. high-tech exports to China" (Pillsbury 2007, 48).

Bruce MacDonald's report on *China, Space Weapons, and U.S. Security* for the Council on Foreign Relations offers a number of noteworthy additional specific recommendations for both the United States and China including: for the United States: assessing the impact of different U.S. and Chinese offensive space postures and policies through intensified analysis and "crisis games," in addition to wargames; evaluating the desirability of a "no first use" pledge for offensive counterspace weapons that have irreversible effects; pursuing selected offensive capabilities meeting important criteria—including effectiveness, reversible effects, and survivability—in a deterrence context to be able to negate adversary space capabilities on a temporary and reversible basis, refraining from further direct ascent ASAT tests and demonstrations as long as China does, unless there is a substantial risk to human health and safety from uncontrolled space object reentry; and entering negotiations on a [kinetic energy] KE-ASAT testing ban. MacDonald's recommendations for China include: providing more transparency in its military space programs; refraining from further direct ascent ASAT tests as long as the United States does; establishing a senior national security coordinating body, equivalent to a Chinese National Security Council; strengthening its leadership's foreign policy understanding by increasing the international affairs training of senior officer candidates and establishing an international security affairs office within the PLA; providing a clear and credible policy and doctrinal context for its 2007 ASAT test and counterspace programs more generally and addressing foreign concerns over China's ASAT test; and offering to engage in dialogue with the United States on mutual space concerns and become actively involved in discussions on establishing international space codes of conduct and confidence-building measures (MacDonald 2008, 34–38).

Finally, Beijing and Washington should pursue specific initiatives to follow up on the cooperative dialogue during the visits of General Xu Caihou and President Obama as well as initiating discussions about recent statements by General Xu Qiliang, Commander of the PLA Air Force (PLAAF), that a space arms race is inevitable and the PLAAF must develop offensive space operations. President Hu quickly repudiated these statements, but the two sides need to find a way to initiate and sustain focused discussions about the difficult space security issues raised by the general's statements since they represent an unprecedented level of public transparency on the part of the PLA, undoubtedly reflect the position of the PLA and other important stakeholders within the Chinese government, and represent an inherent part of the context for space security about which the United States and China must develop better shared understanding. Counterintuitively, Beijing and Washington can lay a stronger foundation for sustainable space security through transparent dialogue over these most difficult issues rather than by trying to avoid them since more diplomatic approaches may assuage but cannot eliminate the growing strategic and military potential of space capabilities.

References

Commission to Assess National Security Space Management and Organization. January 11, 2001. *Report of the Commission to Assess National Security Space Management and Organization.* Washington, DC: Commission to Assess National Security Space Management and Organization.

Gates, R.M. April 17, 2009. Presentation at Naval War College. Newport, RI.

MacDonald, B.W. 2008. *China, Space Weapons, and U.S. Security.* New York: Council on Foreign Relations.

O'Hanlon, M.E. 2004. *Neither Star Wars Nor Sanctuary: Constraining the Military Uses of Space* Washington, DC: Brookings Institution.

Pillsbury, M.P. January 19, 2007. "An Assessment of China's Anti-Satellite and Space Warfare Programs, Policies, and Doctrines." Report prepared for the U.S.-China Economic and Security Review Commission, 48.

Pollpeter, K. March 2008. *Building for the Future: China's Progress in Space Technology During the Tenth 5-Year Plan and the U.S. Response.* Carlisle: Strategic Studies Institute, U.S. Army War College.

Smith, B. 1999. "The Challenge of Space Power." *Airpower Journal* 13(1).

4

Chronology

October 3, 1942	In a successful test flight, the German A-4 ballistic missile (later renamed the V-2 *Vergeltungswaffe* or Vengeance Weapon-2) becomes the first man-made object to brush the edge of space. Major General Walter R. Dornberger and Dr. Wernher von Braun oversee V-2 development at the Peenemünde test site on the Baltic Sea.
May 2, 1946	The newly established research and development (RAND) corporation releases its first report, "Preliminary Design of an Experimental World-Circling Spaceship." The report provides detailed explanation of the technologies and physics involved in building and launching satellites; it also identifies many important potential military missions for satellites including communications, attack assessment, weather reconnaissance, and strategic reconnaissance.
March 1954	President Eisenhower commissions the top secret Technological Capabilities Panel (TCP) and chooses Dr. James R. Killian, President of the Massachusetts Institute of Technology (MIT), as its chairman. Eisenhower established the TCP because he felt it was imperative that the best minds in the country attend to the technological problem of preventing another Pearl Harbor.

August 1954	Air Force Brigadier General Bernard A. Schriever takes command of the newly established Western Development Division (WDD) in El Segundo, California. The WDD was created to expedite Atlas intercontinental ballistic missile (ICBM) development and also initially became responsible for developing the Air Force's first satellite program, the Weapons System (WS)-117L.
November 24, 1954	In an Oval Office meeting, the President, Secretaries of State and Defense, as well as top Department of Defense (DoD) and Central Intelligence Agency (CIA) officials, TCP chairman Killian, and Edwin H. "Din" Land, founder of the Polaroid Corporation and chairman of the intelligence subcommittee of the TCP, discuss the initial programs and structure for a national strategic reconnaissance program including production of a high-altitude reconnaissance jet-powered glider and initiation of America's first space program. The President verbally authorizes the CIA to begin development of the CL-282 (U-2) reconnaissance aircraft at Clarence "Kelly" Johnson's Lockheed "Skunk Works" with Air Force support.
February 1955	The TCP completes a secret two-volume report strongly recommending rapid development of U.S. technical intelligence gathering capabilities and creating policies to support the legality of satellite overflight; the TCP briefs the National Security Council (NSC) on its recommendations.
March 16, 1955	The Air Force sets formal but secret requirements including three-axis stabilization and high-pointing accuracy for the WS-117L reconnaissance satellite program.
March 1955	Eisenhower supports the proposal from the National Academy of Science and the National Science Foundation for an International Geophysical Year (IGY) focused on high-altitude research to be held from July 1, 1957 to December 31, 1958.

Spring 1955   The NSC Planning Board, Special Assistant to the President on Government Operations Nelson Rockefeller, and Assistant Secretary of Defense for Research & Development Donald A. Quarles develop America's first space policy by reviewing and analyzing the differing goals and requirements of the TCP "freedom of space" objective, the WS-117L program, the U.S. IGY satellite proposal, and several military requirements and booster considerations. Quarles produces NSC-5520, "Draft Statement of Policy on U.S. Scientific Satellite Program," a secret document the NSC approves on May 26, 1955, and the President signs the next day.

July 1955   White House Press Secretary James Hagerty publicly announces that the President has approved plans for launching small, Earth-circling satellites as part of U.S. participation in the IGY.

Each of the services makes presentations to the Stewart Committee and competes for the honor of having its booster open the space age by launching the U.S. IGY satellite.

August 3, 1955   The Stewart Committee votes 3–2 in favor of the Naval Research Laboratory's (NRL) proposal to develop an upgraded version of the Navy's Viking sounding rocket to launch the U.S. IGY satellite under Project Vanguard.

September 20, 1956   A Jupiter-C missile developed by the Army Ballistic Missile Agency (ABMA) under Major General John B. Medaris and Dr. von Braun establishes new records by reaching 600 miles high and traveling 3,000 miles downrange. The missile carries an inert fourth stage "kick motor" with sand as ballast weight lest it "accidentally" place a satellite into orbit prior to Project Vanguard.

October 4, 1957   The Soviet Union opens the space age by launching *Sputnik* I.

October 8, 1957	Eisenhower privately calls Quarles to the White House to discuss the space-policy goals of his administration. Quarles explains that the Russians have unintentionally helped the United States by establishing the concept of freedom of space. Eisenhower agrees to Quarles's request to study using the Jupiter-C as a quicker and surer way to launch America's first satellite.
October 15, 1957	Eisenhower meets with the distinguished scientists who make up the Science Advisory Committee and decides to appoint James Killian to the new position of Special Assistant to the President for Science and Technology.
November 1957– January 1958	Senate majority leader Lyndon Johnson holds a series of hearings that provide an important forum for top civilian and military leaders to express their views about the strategic importance of space and the state of U.S. national security.
November 1957	Eisenhower and the NSC receive the Gaither Report from a top secret strategic review committee commissioned in the spring. Chairman Rowan Gaither presents a very somber picture of the current and future strategic balance between the superpowers and is especially concerned with the vulnerability of the U.S. bomber force in the space and missile age.
November 3, 1957	The Soviets send into orbit the 1,121-pound *Sputnik* II with *Laika* the dog aboard.
November 15, 1957	Secretary of Defense Neil McElroy announces creation of the Advanced Research Project Agency (ARPA) to consolidate and control all DoD research and development (R&D) efforts including space programs.
November 29, 1957	Air Force Chief of Staff General Thomas D. White provides the first comprehensive and official expression of Air Force space doctrine in a presentation at the National Press Club.

December 1957	The President's Science Advisory Committee (PSAC) emerges at a consensus that scientifically oriented civil space missions ought to be the nation's top space priority and that a civilian space agency built from and modeled after the National Advisory Committee on Aeronautics (NACA) would provide the best organizational approach.
December 6, 1957	The launch pad explosion of Vanguard TV-3 is televised and showcases the spectacular failure of America's first satellite-launch attempt worldwide.
January 1958	Eisenhower initiates a series of exchanges with Soviet Premier Nikolai Bulganin in a letter proposing the superpowers agree to use space for "peaceful purposes" only.
January 31, 1958	An Army Juno (Jupiter-C) successfully launches *Explorer* I, America's first satellite, from Cape Canaveral. The scientific instruments aboard the satellite designed by Dr. James Van Allen detect the radiation belts around Earth that are later named in his honor.
February 1958	ARPA assumes control of the WS-117L program and this becomes the agency's single most important space project, accounting for $152 million or nearly one-third of ARPA's 1958 budget.
February 3, 1958	Eisenhower directs that development of ICBMs, intermediate range ballistic missiles (IRBMs) and the WS-117L satellite programs be given highest and equal national priority.
February 3–4, 1958	The PSAC is formally tasked to study space mission priorities and recommend possible organizational structures. The next day, this PSAC study, which came to be known as the Purcell Report after its chairman Edward Purcell of Harvard, is initiated and publicly announced.
February 4, 1958	During a private meeting with top Republican leaders in Congress, Eisenhower indicates his strong preference for keeping civil space efforts

within ARPA in order to keep his top-priority WS-117L program shielded and on track while avoiding the duplication he saw arising from creation of a civil space agency.

February 7, 1958

In a meeting with Killian and Land, Eisenhower makes the CIA, rather than the Air Force, primarily responsible for development of a reconnaissance satellite using the recoverable film method (Corona Program) and schedules it to be operational by the spring of 1959.

March 5, 1958

Based on a Killian memorandum on the recommendations of the Purcell Committee, Eisenhower changes his position and responds enthusiastically to the plan to create a civilian space agency.

April 1958

The Air Force releases a plan to obtain approval for its ambitious manned space program prior to the establishment of a civilian space agency.

April 2, 1958

Killian supervises drafting of proposed legislation on creation of a civilian space agency and delivers it to Congress.

July 7, 1958

Eisenhower and Senate majority leader Lyndon Johnson resolve a deadlock between the House and Senate versions of legislation to create a civilian space agency. Eisenhower agrees to accept Johnson's NSC-type Space Council at the White House if the President is made chairman of the committee.

July 15, 1958

A conference committee works out final compromises on the National Aeronautics and Space Act of 1958 including a modified version of the House's Civilian-Military Liaison Committee between the National Aeronautics and Space Administration (NASA) and DoD, a National Aeronautics and Space Council (NASC) at the White House, and the final wording of Section 2 indicating NASA would exercise control over

U.S. space activities "except that activities peculiar to or primarily associated with the development of weapons systems, military operations, or the defense of the U.S. [including the Research and Development necessary to make effective provision for the defense of the U.S.] shall be the responsibility of and shall be directed by the DOD."

July 29, 1958   Eisenhower signs the National Aeronautics and Space Act of 1958 creating NASA.

August 1958   Quarles and Killian reach a compromise that directs NASA to design and build the capsules for manned spaceflight and ARPA to concentrate on the boosters required for this mission.

Eisenhower formally gives NASA primary authority over U.S. manned spaceflight efforts and this effort evolves into Project Mercury.

August 18, 1958   The United States adopts its first comprehensive space policy in a secret document, NSC 5814/1, "Statement of Preliminary U.S. Policy on Outer Space."

August–September 1958   The United States initiates a series of high-altitude nuclear detonations that are designed to test and do confirm that high-energy electrons produced in such detonations become trapped in Earth's magnetic field.

October 1, 1958   NASA is established.

October 1958   Quarles and Keith Glennan, NASA's first Administrator, try to resolve NASA's lack of specific space expertise, especially in booster development, by transferring the Army-sponsored Jet Propulsion Laboratory (JPL) at the California Institute of Technology and the von Braun team at ABMA to NASA. ABMA Commander Medaris and Army Secretary Wilber Brucker vigorously fight against this proposal and the NASC has to work out a compromise allowing JPL to become part of NASA and the von Braun team to remain under ABMA but work on the Saturn booster under contract to NASA.

November 24, 1958
The United Nations Ad Hoc Committee on the Peaceful Uses of Outer Space (COPUOS) is created.

February 28, 1959–June 1960
The Corona program begins with the liftoff of *Discoverer* I from Vandenberg Air Force Base (AFB), California, but significant technological reliability problems with the Thor-Agena launch vehicles as well as various control glitches with the satellites themselves and with the film ejection and recovery system prevent any successful film retrievals from the first 12 launches.

March 1959
In an attempt to find a better rationale to promote their space capabilities, the Army organizes Project Horizon, a study of military uses for the Saturn booster.

Summer 1959
The Army and Navy propose creation of a unified or joint-service space command with the responsibility for development and production of all space vehicles and boosters. The Chief of Naval Operations (CNO), Admiral Arleigh Burke, forwards a unified space command proposal from the Joint Chiefs of Staff (JCS) to Secretary of Defense McElroy.

June 1959
The Project Horizon Report is completed and details a comprehensive plan to establish a 12-man lunar outpost by November 1966; construction of the lunar outpost is to be supported by 149 Saturn launchings and is estimated to cost $6 billion.

September 18, 1959
Secretary of Defense McElroy rejects Burke's proposal for a unified space command in a memorandum transferring satellite programs from ARPA and making the Air Force responsible for almost all DoD booster development programs and space launches.

October 19, 1959	Under Project Bold Orion, the United States conducts the world's first successful ASAT test by air-launching a missile from B-47 bomber that passes within four miles of the *Explorer* VI satellite.
January 1960	Congress approves ABMA's realignment under NASA, a transfer supported by Science Adviser George Kistiakowsky, Quarles, DDR&E York, and Eisenhower.
May 1960	Eisenhower directs Kistiakowsky to study what corrective actions might be necessary to achieve success in developing reconnaissance satellites. Kistiakowsky and Defense Secretary Thomas Gates decide on a study committee known as the Samos Panel comprised of Under Secretary of the Air Force Joseph Charyk, Deputy DDR&E John Rubel, and Kistiakowsky.
June 22, 1960	The secret Galactic Radiation and Background (GRAB) electronic intelligence collection payload is launched by NRL and becomes the world's first intelligence gathering satellite (spysat).
July 1, 1960	Eisenhower presides over the opening of NASA's Marshall Space Flight Center in Huntsville, Alabama.
August 20, 1960	Corona becomes the world's first operational satellite photoreconnaissance system when the film canister ejected from *Discoverer* XIV is successfully retrieved in midair by a C-119.
August 25, 1960	The Samos Panel recommends to the NSC immediate creation of an organization to provide a direct chain of command from the Secretary of the Air Force to the officers in charge of each reconnaissance satellite project. This recommendation leads to creation of the National Reconnaissance Office (NRO) a year later.
October 1960	General Schriever asks Trevor Gardner to chair the Air Force Space Study Committee and examine future military options in space.

January 1961	Science Adviser Jerome Wiesner's Ad Hoc Committee on Space recommends revitalizing the NASC, placing primary emphasis on space-science missions, avoiding races with the Soviets for manned space spectaculars, and preventing DoD's overlapping space programs and duplication of NASA's work. Secretary of Defense Robert S. McNamara agrees with the tenor of these recommendations and tasks his office to begin the review of military space organizations that leads to Defense Directive 5160.32, "Development of Space Systems."
March 1961	The top secret Gardner Report is completed and provides a ringing endorsement of the high-ground and space-control schools already prevalent within the Air Force. The report basically ignores NASA and calls for a new Air Force Systems Command (AFSC) to spearhead an accelerated and very ambitious program including manned spaceflight, space weapons, reconnaissance systems, large boosters, space stations, and even a lunar landing by 1967–1970.
March 6, 1961	McNamara signs DoD Directive 5160.32, "Development of Space Systems," making the Air Force responsible for developing and launching almost all DoD space systems.
April 12, 1961	The Soviet Union's Yuri Gagarin completes the first successful manned orbital flight.
August 1961	The highly classified NRO is formally established; its very existence is an official U.S. state secret until September 1992.
September 1961	The first classified National Intelligence Estimate (NIE) to incorporate spysat data is completed and indicates the United States believes the Soviet Union has fewer than 10 operational ICBMs, quieting fears of a missile gap.
December 20, 1961	The UN passes General Assembly (UNGA) Resolution 1721, requiring all states to provide data on their space launches to the UN.

March 23, 1962	Secret DoD Directive S-5200.13, "Security and Public Information Policy for Military Space Programs," is issued. Also known as the "Blackout" Directive, this is a final step in a security clampdown on all military space efforts that prohibits advance announcement and press coverage of military space launches, forbids use of the names of military space programs such as Midas and Samos, and replaces program names with program numbers.
May 26, 1962	President John F. Kennedy issues National Security Action Memorandum (NSAM) 156—an implicit critique of the NASC and a request for the Department of State to create a high-level coordinating body (known as the NASM 156 Committee) to address a lack of interagency coordination.
Summer–Fall 1962	The United States conducts another series of high-altitude nuclear detonations including the July 9 1.4-megaton Starfish Prime test 248 miles above Johnson Island in the Pacific that caused a large electromagnetic pulse resulting in electrical problems in Hawaii, almost 900 miles away, and failure of seven operational satellites in low-Earth orbit over the next several weeks.
	The NSAM 156 Committee is the scene of intense interagency disputes over banning nuclear weapons from space with the State Department and the Arms Control and Disarmament Agency (ACDA) most supportive of a ban and the JCS most strongly opposed.
December 1962	The SAINT program, an Air Force proposal for a satellite interception and inspection system, the largest and most comprehensive initial ASAT R&D effort, is canceled before any development or testing.
December 3, 1962	In a key speech at the UN, Ambassador Albert Gore asserts that: "It is the view of the United

States that Outer Space should be used for peaceful—that is non-aggressive and beneficial—purposes."

January 1963 NASA Administrator James Webb and McNamara sign an agreement to allow DoD experiments on some Gemini missions in a program known as Blue Gemini.

August 1963 The Army's Program 505, a modified Nike Zeus missile stationed at Kwajalein Atoll, becomes operational. This is first of two nuclear-tipped ASAT systems the United States deploys in the Pacific.

October and December 1963 The UN passes UNGA Resolutions 1884 and 1962: an international declaratory ban on placing nuclear weapons or weapons of mass destruction in outer space and a declaration that space is to be free for exploration by all, will not be subject to appropriation by national sovereignty, and that launching states will maintain jurisdiction over spacecraft and are liable for damage they might cause.

December 10, 1963 Secretary McNamara cancels the X-20 or dynamic soaring proposal, ending the early Air Force manned military spaceflight dreams and at the same time continues the building block approach to determine the military utility of human spaceflight by assigning primary responsibility for developing the Manned Orbiting Laboratory (MOL) to the Air Force.

June 1964 The Air Force's Program 437 Thor ASAT, located on Johnson Island, becomes operational and is the second nuclear-tipped ASAT system the United States deploys.

April 5, 1966 National Security Adviser Walt Rostow writes a memorandum to President Johnson recommending the United States rapidly propose an international treaty at the UN to codify the principles in UNGA 1962 before the Soviets table their own draft treaty on space-related arms control.

May 7, 1966	President Johnson publicly outlines the basic provisions of the U.S. draft treaty on space-related arms control.
December 17, 1966	Following negotiations and adoption of some provisions from the Soviet draft treaty, the UNGA endorses the agreement on space-related arms control.
January 27, 1967	The Treaty on Principles Governing the Activities of States in the Exploration and Use of Outer Space, Including the Moon and Other Celestial Bodies (Outer Space Treaty or OST) becomes open for signature. Sixty-two states initially sign the OST.
February 7, 1967	The OST goes to the U.S. Senate for advice and consent for ratification.
March 1967	President Johnson's off-the-record yet widely cited remarks to a group of Tennessee educators refer directly back to the missile-gap episode and reflect his faith in and enthusiasm for space-based reconnaissance.
April 25, 1967	The Senate votes 88–0 in support of ratification of the OST.
October 1968– December 1971	The Soviets conduct seven tests of a co-orbital ASAT system that is launched by a modified SS-9 ICBM booster and employs a radar-guided explosive warhead.
February 1969	Incoming Secretary of Defense Melvin Laird endorses a comprehensive review of the MOL program that supports the program and finds it can meet all its objectives with only six launches (two unmanned and four manned).
June 10, 1969	Budgetary pressures and the ongoing war in Vietnam prompt the Nixon Administration to cancel the MOL program.
July 20, 1969	*Apollo* 11 astronauts Neil Armstrong and Buzz Aldrin become the first men to walk on the Moon.

March 1970   President Richard M. Nixon formalizes U.S. post-Apollo space-policy goals by endorsing just half of the least ambitious recommendation from his Space Task Group: development of a space shuttle; he leaves a space station or Mars mission contingent upon successful completion of a shuttle program.

January 5,
1972   The Air Force sets a number of challenging performance criteria including a requirement to lift a 65,000-pound payload and moves NASA toward a lifting body design known as the Thrust-Assisted Orbiter Shuttle (TAOS) that is formally approved as the Space Transportation System (STS) by Nixon.

May 26,
1972   SALT I consisting of the Anti-Ballistic Missile (ABM) Treaty and the Interim Agreement on the Limitation of Strategic Offensive Arms is signed and heralds the arrival of the era of detente between the superpowers. SALT I is enabled by spysats (euphemistically referred to as National Technical Means or NTM of verification) and is an attempt to codify mutual assured destruction (MAD) as the putative basis for strategic stability.

February
1976   The Soviets resume testing their co-orbital ASAT system using new guidance seekers, beginning a 13-launch test series lasting until June 1982.

Fall 1976   President Gerald Ford issues National Security Decision Memorandum (NSDM)-333 directing DoD to reduce satellite vulnerability problems through increased effort and funding.

December
19, 1976   First launch of the revolutionary fifth-generation electro-optical spysat known as the KH-11; the filmless system solves the timeliness problem with a near-real time direct data downlink.

January 18,
1977   Ford issues NSDM-345 authorizing a new non-nuclear ASAT, leading in September to initiation

of the Air Force's air-launched miniature homing vehicle (MHV) program.

May 11, 1978   President Jimmy Carter issues secret Presidential Directive (PD)-37, "National Space Policy." The directive's main provisions include working to resolve potential conflicts among space program sectors; using the STS for all authorized users, domestic, foreign, commercial and governmental; and an aggressive, long-term program to provide greater survivability for military space systems.

June 8–16, 1978   U.S. and Soviet negotiators met in Helsinki for ASAT talks.

October 10, 1978   National Security Adviser Zbigniew Brzezinski signs PD-42, "Civil and Further National Space Policy," indicating three components for U.S. civil space policy: (1) evolutionary space activities unique to or more efficiently accomplished in space; (2) a balanced strategy of applications, science, and technology development; and (3) emphasis that it was neither feasible nor necessary to initiate a challenging space program comparable to Apollo at this time.

January 23 – February 16, 1979   U.S. and Soviet negotiators meet in Bern for ASAT talks.

April 23– June 17, 1979   U.S. and Soviet negotiators meet in Vienna for ASAT talks.

April 12, 1981   The initial spaceflight of *Columbia* marks a bittersweet milestone because it is the world's first reusable spacecraft and signifies the return of American human spaceflight, but the STS is also two years behind schedule and costs $2 billion more to develop than originally projected.

July 4, 1982   President Ronald Reagan publicly announces his administration's first major space policy, National Security Decision Directive (NSDD)-42.

The directive, written by the Senior Interagency Group or SIG (Space) and chaired by Science Adviser George Keyworth, is far less enthusiastic about space-related arms control than the Carter Administration had been.

September 1, 1982 Air Force Space Command (AFSPC), the first completely new major command formed by the Air Force in 32 years, is created. AFSPC is designed to consolidate, centralize, and focus many of the Air Force's space efforts.

February 11, 1983 Faced with difficulties in modernizing U.S. strategic nuclear forces, particularly in finding a suitable basing mode for the new MX "Peacekeeper" ICBM; growing strength of the nuclear freeze movement worldwide; and continuing Soviet strategic modernization, Reagan moves strongly toward strategic defenses after receiving unanimous support for investigating new defense possibilities at a White House meeting with Secretary of Defense Caspar Weinberger, National Security Adviser "Bud" McFarlane, and the JCS.

March 23, 1983 President Reagan's "Star Wars" speech begins the Strategic Defense Initiative (SDI).

July 18, 1983 Senator Paul Tsongas's amendment to the Fiscal Year (FY) 1984 DoD Authorization Act withholds funds for testing the MHV ASAT system unless the President certifies both that the United States was negotiating with the Soviets in good faith about ASATs and that testing is needed for U.S. national security.

August 19, 1983 Soviet General Secretary Yuri V. Andropov announces a unilateral moratorium on ASAT testing so long as other countries refrain from testing or stationing ASATs in space.

January 1984 The Air Force's MHV ASAT program is first flight tested.

January 6,
1984

Two major studies, the "Defensive Technologies Study," or Fletcher study after its chairman, former (and future) NASA Administrator James Fletcher; and the "Future Security Strategy Study" undertaken by two groups, an interagency team led by Franklin Miller and a team of outside experts chaired by Fred Hoffman, find that many emerging technologies hold significant missile-defense potential. They recommend building defenses in multiple layers, including space-based layers, and provide strong support for the formal authorization to begin SDI and establish the SDI Office (SDIO) reporting to the Secretary of Defense.

March 31,
1984

As directed, the administration delivers a report, "U.S. Policy on ASAT Arms Control" to Congress. The report raises several questions concerning the basic strategic utility of an ASAT ban and strongly reiterates the administration's many concerns, noting in particular that "deterrence provided by a U.S. ASAT capability would inhibit Soviet attacks against U.S. satellites, but deterrence is not sufficient to protect U.S. satellites. Because of the potential for covert development of ASAT capabilities and because of the existence of non-specialized weapons which also have ASAT capability, no arms control measures have been identified which can fully protect U.S. satellites. Hence, we must continue to pursue satellite survivability measures to cope with both known and technologically possible, yet undetected, threats."

April 15,
1984

Lieutenant General James Abrahamson moves from his position as Associate Administrator for Manned Spaceflight at NASA to become the first director of SDIO.

June 10,
1984

SDIO demonstrates the potential of emerging defense technologies when a kinetic energy interceptor known as the Homing Overlay Experiment launches from Meck Island in the Pacific

Test Range and successfully intercepts a test reentry vehicle launched atop a Minuteman ICBM from Vandenberg AFB.

February 20, 1985   Special Arms Control Ambassador Paul Nitze officially codifies two criteria by which SDI developments should be judged: first, that any defense systems be highly survivable, and second, that defense systems be cost effective at the margin—that is, the incremental cost of adding additional defensive capability must be low enough so that the other side has little incentive to add additional offensive capability to overcome the defense.

March 1985   ASAT negotiations are restarted as a subset of the broad Defense and Space Talks.

September 13, 1985   Based on resumption of negotiations and continuing administration pressure, congressional restrictions are relaxed and the MHV successfully performs its first and only intercept against a satellite in space. The MHV is launched by an F-15 from Edwards AFB, California, and destroys the Solwind (P-78-1) satellite at an altitude of 248 miles; the last of the 285 pieces of debris created by this test that are large enough to be tracked take 17 years to reenter Earth's atmosphere and burn up.

September 23, 1985   A unified command structure consisting of all services, United States Space Command (USSPACECOM), is established above AFSPC, but is primarily funded by the Air Force and staffed by Air Force personnel.

January 28, 1986   The *Challenger* disaster occurs and grounds all planned national security launches for more than two years.

September 5, 1986   A series of SDIO experiments, known as the Delta 180 test, confirms the ability of space-based infrared sensors and kinetic weapons to perform simulated boost-phase intercepts.

December 27, 1986	NSDD-254, "United States Space Launch Strategy," is completed and indicates the United States would rely upon a balanced mix of launchers consisting of the STS and expendable launch vehicles (ELVs) and that critical payloads would be designed for dual-compatibility, i.e., launch by either STS or ELV.
May 1987	Partially due to missile-defense requirements that might require a heavy lift capability, DoD initiates a joint program with NASA to develop a new ELV program known as the Advanced Launch System (ALS). The goals of the ALS program include the development of a flexible and reliable family of modular launch vehicles that can easily be configured for specific needs.
September 18, 1987	The JCS establishes classified minimum performance requirements for a Phase I strategic defense deployment and Secretary Weinberger announces that the Defense Acquisition Board had formally moved six parts of the Phase I SDI program past the demonstration and validation milestone of the defense acquisition process.
December 1987	The Air Force cancels the MHV after two years of additional funding cutbacks and a congressional prohibition against testing unless the Soviets first performed a dedicated test of their co-orbital system.
January 5, 1988	Due to major developments including the Strategic Defense Initiative (SDI) and *Challenger* disaster, the Reagan Administration issues a revised comprehensive and secret directive on overall U.S. national space policy. The White House fact sheet released on February 11 indicates Reagan's new policy again exhibits great continuity with earlier space policies but does include some significant changes: maintenance of U.S. space leadership is no longer a basic goal; more emphasis is placed on promoting commercial space activities;

and expanding human presence and activity beyond Earth orbit is added as a basic goal.

November 2, 1989 The newly created National Space Council, chaired by Vice President Dan Quayle, develops a classified national space-policy directive for President George H. W. Bush. On November 18, the Space Council releases the unclassified "National Space Policy," which is virtually identical to the February 11, 1988, fact sheet on the last Reagan space policy.

December 1989 The Army becomes the executive agent for developing a new ground-launched ASAT system known as the Kinetic Energy ASAT (KE-ASAT).

September 1990 Bush continues the policy of launching U.S. government satellites only on U.S. manufactured launch vehicles.

Fall 1990 The U.S. moves the two newest and most capable DSP satellites into GEO overhead the Persian Gulf region to provide stereo imaging of missile launches from this area.

February 1991 NSPD-3, "U.S. Commercial Space Policy Guidelines," directs U.S. government agencies to use commercial space products, to encourage development of this sector, maintain U.S. space preeminence, and save money in areas including: satellite communications, launch services, remote sensing, material processing, and development of commercial infrastructure.

July 10, 1991 NSPD-4, "National Space Launch Strategy," supports assured access through a mixed fleet of Shuttles and ELVs and directs that DoD and NASA jointly undertake development of a new medium- to heavy-lift vehicle to reduce launch costs while improving reliability and responsiveness.

February–March 1992 President Bush issues NSPD-5 and -6, on Landsat Remote Sensing Strategy and the Space Exploration Initiative.

Fall–Winter 1992	The goal of building and testing a full-size, manned National Aerospace Plane (NASP or X-30) is abandoned. The program was designed to build an experimental single-stage-to-orbit vehicle that would take off like an airplane, fly into space, and then return to land like an airplane but proves to be extremely challenging technologically and fiscally. A December 1992 Government Accountability Office (GAO) report estimates the total costs for a complete X-30 program would have been $17 billion versus the $3.1 billion original estimate in the late 1980s.
September 1993	Vice President Al Gore announces that the Russian Federation will join the International Space Station (ISS) effort and also abide by the terms of the Missile Technology Control Regime (MTCR).
March 1994	President William J. Clinton issues policies on Foreign Access to Remote Sensing Space Capabilities in Presidential Decision Directive in (PDD)-23.
May 1994	Clinton issues policies on Convergence of U.S.-Polar Orbiting Environmental Satellite Systems in NSTC-2.
	Clinton issues policies on Landsat Remote Sensing Strategy in NSTC-3.
August 1994	Clinton issues policies on National Space Transportation Policy in NSTC-4.
March 29, 1996	NSTC-6, "U.S. Global Position System Policy," is released and indicates the United States will continue to provide GPS service without direct user fees, encourage worldwide acceptance and integration of GPS, and discontinue deliberately degrading the accuracy of signals provided to non-military users (Selective Availability) within a decade.
September 19, 1996	The White House releases PDD-49, "National Space Policy," which reemphasizes the United States will maintain a leadership role in space by

supporting a strong, stable, and balanced program across all space-activity sectors.

January 11, 2001

The congressionally mandated Commission to Assess United States National Security Space Management and Organization, generally known as the Rumsfeld Space Commission after it chairman and future Secretary of Defense Donald H. Rumsfeld delivers a report that remains the single most comprehensive and influential examination of NSS issues.

October 1, 2002

USSPACECOM is merged into United States Strategic Command (USSTRATCOM).

April 25, 2003

"U.S. Commercial Remote Sensing Space Policy" summarizes National Security Presidential Directive (NSPD)-27, the classified policy that supersedes PDD-23, indicating that the U.S. government will competitively outsource functions in order to rely on U.S. commercial remote-sensing capabilities for filling its imagery and geospatial needs; focus its remote-sensing capabilities on needs that cannot be satisfied by commercial providers; and maintain a Sensitive Technology List that would be approved for export only rarely, on a case-by-case basis.

January 2004

President Bush announces his "Vision for Space Exploration" that seeks to advance U.S. scientific, security, and economic interests through a robust space-exploration program including a sustained and affordable human and robotic program to explore the solar system and beyond, starting with a human return to the Moon by the year 2020, in preparation for human exploration of Mars and other destinations.

December 15, 2004

"U.S. Space-Based Position, Navigation, and Timing Policy" explains the policy that superseded NSTC-6 and noted that GPS was now a part of the worldwide economy.

January 6, 2005

The Bush Administration releases "U.S. Space Transportation Policy," explaining the policy that supersedes NSTC-4 including goals to: demonstrate initial capabilities for operationally responsive access to and use of space; develop capabilities for human exploration beyond LEO; and sustain and promote a domestic space-transportation industrial base.

August 31, 2006

NSPD-49, "U.S. National Space Policy," is signed by George W. Bush. The policy deals with most of the same space issue areas as the Clinton National Space Policy 10 years earlier but takes a slightly more competitive tone.

October 2009

The Review of U.S. Human Spaceflight Plans Committee (Augustine Committee) reports that implementation of the Constellation Program, developed by NASA Administrator Michael Griffin to realize President Bush's "Vision for Space Exploration," will take an additional $3 billion annually.

February 2010

The Obama Administration does not request funding for the Constellation Program in the NASA budget submitted to Congress. The budget request also terminates the National Polar-Orbiting Operational Environmental Satellite System (NPOESS) by reverting to separate environmental monitoring systems.

June 28, 2010

The Obama Administration releases "The National Space Policy of the United States of America."

5

Biographical Sketches

James A. Abrahamson (May 19, 1933–)

Air Force Lieutenant General James Abrahamson served as Director of the Strategic Defense Initiative Office (SDIO) from 1984 to 1989. He was an energetic, positive, and sales-oriented advocate for President Reagan's original comprehensive vision of SDI. Before his appointment at SDIO, he was the National Aeronautics and Space Administration (NASA) Associate Administrator for Manned Space Flight; on April 15, 1984, Abrahamson moved from NASA to become the first director of SDIO. As the Associate Director of Manned Space Flight he emphasized the role of the Space Transportation System (STS or Space Shuttle) in facilitating military missions, including the potential to develop manned military space stations. Abrahamson and the SDIO first demonstrated the potential of new ballistic missile-defense (BMD) technologies on June 10, 1984, when a kinetic energy weapon (KEW) known as the Homing Overlay Experiment (HOE) launched from Meck Island in the Pacific Test Range successfully intercepted a test reentry vehicle (RV) launched atop a Minuteman intercontinental ballistic missile (ICBM) from Vandenberg Air Force Base (AFB). Later, Abrahamson pushed for SDIO to absorb Defense Advanced Research Projects Agency (DARPA) space laser experiments into its Space-Based-Laser (SBL) directed energy weapons (DEW) program element. None of these space laser programs were tested in space during Abrahamson's time as SDIO Director. Despite a long string of successful assignments and being perceived as a rising star within the Air Force prior to becoming director of SDIO, Abrahamson's nomination by President Reagan for full general was blocked by forces within the Department of Defense (DoD) and the Senate because they were concerned his promotion would increase the influence of SDIO in Pentagon decision making. Abrahamson retired at the end of January 1989 without making full general.

Edward C. Aldridge ("Pete") (August 18, 1938–)

Pete Aldridge served as Under Secretary of the Air Force from 1981 to 1986, as Secretary of the Air Force from 1986 to 1988, and was Director of the National Reconnaissance Office (NRO) from 1981 to 1988. He trained to be a payload specialist on the first STS scheduled to be launched from Vandenberg AFB (VAFB) in 1986 but this mission was canceled following the *Challenger* disaster; Aldridge never flew in space and no shuttles ever launched from VAFB. Aldridge supported the Air Force's creation in 1982 of a separate major command for space. To promote efficiency and joint operational effectiveness he argued that DoD needed the Air Force to continue serving as single manager for many DoD space activities, including for multiuser programs such as the Global Positioning System (GPS), Defense Metrological Support Program (DMSP), and Milstar. Despite this position, Aldridge also strongly supported creation in 1985 of a unified space command, arguing that was the next logical step for military space organizations and that a unified space command was necessary for the future. During 1983 and 1984, Aldridge waged a mostly secret and very difficult but eventually successful campaign against NASA to obtain approval to develop a new expendable launch vehicle (ELV) capable of launching spy satellites designed to fit into the STS. In December 1983, Aldridge issued a memorandum on assured access to space to Air Force Systems Command and Air Force Space Command directing these organizations to plan for procurement of a complementary ELV (CELV) capable of boosting a payload the size of the STS cargo bay and weighing 10,000 pounds into geostationary Earth orbit (GEO). In August 1984, the National Security Council (NSC) formally supported Aldridge's position by issuing National Security Decision Directive (NSDD)-144 approving Air Force development of the CELV. Next, Aldridge was instrumental in persuading the administration and Congress to authorize the Titan IV (formerly the Titan 34D-7) CELV program in 1985; then, in April

1987, he pushed for development of a new national launch system, a vehicle that would be larger than the STS or Titan IV. Aldridge's goals in developing the CELV were to provide a backup launch capability for the STS; make the U.S. space-launch infrastructure more robust, capable, and cost-effective; and eventually make ELVs rather than the STS the primary launch vehicle for the Air Force. Aldridge was sometimes criticized for allowing budget constraints rather than strategy to dictate the level of Air Force commitment to space. In responding to these criticisms, he indicated that the military was becoming more aware of just how valuable space systems had become to all terrestrial operations and cited the Air Force's long-term funding of over $10 billion to GPS as a specific example of the depth of Air Force commitment to space.

Norman R. Augustine (July 27, 1935–)

Norm Augustine rose through the ranks in a variety of aerospace companies to become chief executive officer (CEO) of Martin Marietta in 1987 and then served as the first President and Chairman of the Lockheed Martin Corporation from 1995 to 1997. He is also very active in a variety of commissions and volunteer work; for example, he served on the President's Council of Advisers on Science and Technology in both Democratic and Republican administrations and chaired the National Academy of Engineering from 1994 to 1996. In 1990, he chaired the Advisory Committee on the Future of the United States Space Program; the report from this committee rank ordered five space objectives and found the U.S. government should give most priority to the advancement of space science. The committee ranked the next priorities as development of space technologies, the study of Earth from space, development of unmanned launch vehicles, and finally manned space flight; it reevaluated these priorities because of fears that lower-ranked objectives would lose funding, reorganizing the objectives so that space science still remained the top priority, but all others were of equal importance as the second objective. In May 2009, Augustine was named as chairman of the Review of U.S. Human Space Flight Plans Committee, tasked to assess NASA's plans for the Moon, Mars, and beyond. The committee's report released in October 2009 found that it would take an additional $3 billion annually to support current human spaceflight plans, did not provide support for NASA's current hardware development under the Constellation program, and was used by the Obama Administration to terminate Constellation funding in their budget request sent to Congress in February 2010.

James M. Beggs (January 9, 1926–)

James Beggs was director of NASA from 1981 to 1985. In 1983, Beggs and Deputy Administrator Hans Mark privately agreed that NASA should push a permanently manned space station as the nation's next major civil space goal, and this major effort became Beggs's focus and legacy. President Reagan announced initiation of Space Station Freedom in his January 1984 State of the Union Address. In the launch vehicle area, Beggs opposed Air Force Under Secretary Aldridge's push to create a new ELV, worrying that the Air Force was attempting to abandon NASA's space shuttle. After Aldridge's position was formally supported by the NSC in August 1984 and the CELV program began, Beggs and NASA continued to contest the CELV option and enlisted considerable congressional support in this opposition.

George H. W. Bush (June 12, 1924–)

George H.W. Bush served as Director of Central Intelligence (DCI) from 1976 to 1977 and President of the United States from 1989 to 1993. As DCI, Bush approved initiating development of the first U.S. radar imaging satellites in 1976. The Bush Administration continued the Reagan Administration's emphasis on space and issued a number of National Space Policy Directives (NSPDs). This was the first administration since Johnson's to have a formal, standing White House organization, the National Space Council (NSpC), to coordinate high-level space issues and develop national space policy. President Bush signed his first space-policy directive on November 2, 1989, and a White House fact sheet was distributed on November 16; this policy was nearly identical to the last space policy of the Reagan Administration. In September 1990, Bush continued the policy of launching U.S. government satellites only on U.S. manufactured launch vehicles and in February 1991 in NSPD-3,

"U.S. Commercial Space Policy Guidelines," directed U.S. government agencies to use commercial space products to encourage development of this sector, maintain U.S. space preeminence, and save money in areas including: satellite communications, launch services, remote sensing, material processing, and development of commercial infrastructure. NSPD-4, "National Space Launch Strategy," issued on July 10, 1991, supported assured access through a mixed fleet of shuttles and ELVs and directed that DoD and NASA jointly undertake development of a new medium- to heavy-lift vehicle to reduce launch costs while improving reliability and responsiveness. In February and March 1992, President Bush issued NSPD-5 and -6, on Landsat Remote Sensing Strategy and the Space Exploration Initiative.

George W. Bush (July 6, 1946–)

George W. Bush served as President of the United States from 2001 to 2009. On April 25, 2003, OSTP released a fact sheet on "U.S. Commercial Remote Sensing Space Policy" indicating that the U.S. government would: competitively outsource functions in order to rely on U.S. commercial remote-sensing capabilities for filling its imagery and geospatial needs; focus its remote-sensing capabilities on needs that cannot be satisfied by commercial providers; and maintain a Sensitive Technology List that would be approved for export only rarely, on a case-by-case basis. In January 2004, President Bush announced his "Vision for Space Exploration" that sought to advance U.S. scientific, security, and economic interests through a robust space-exploration program, including a sustained and affordable human and robotic program to explore the solar system and beyond, starting with a human return to the Moon by the year 2020, in preparation for human exploration of Mars and other destinations. On January 6, 2005, the Bush Administration released "U.S. Space Transportation Policy" including goals to: demonstrate initial capabilities for operationally responsive access to and use of space; develop capabilities for human exploration beyond LEO; and sustain and promote a domestic space-transportation industrial base. The policy also made the Secretary of Defense responsible for maintaining two Evolved ELV (EELV) providers, called for the Secretary to reevaluate this policy with the Director of National Intelligence and NASA Administrator no later than 2010, and charged the NASA Administrator to develop options to meet potential exploration-unique requirements for heavy lift beyond the capabilities of the EELV that emphasize the potential for using EELV derivatives as well as evaluations of the comparative costs and benefits of a new dedicated heavy-lift launch vehicle or options based on shuttle-derived systems. The Bush Administration tied many space issues together in a classified National Space Policy, Presidential Decision Directive (PDD)-49, that was signed on August 31, 2006, and a public fact sheet was released on October 6. This policy was criticized worldwide for having a bellicose tone, but it had great continuity with previous national space policies, reemphasizing the United States would maintain a leadership role in space by supporting a strong, stable, and balanced program across all space-activity sectors. Finally, on September 19, 2007, Bush's White House released a fact sheet announcing that future GPS systems would no longer acquire the capability to degrade signals to nonmilitary users.

Stephen A. Cambone (June 22, 1952–)

Dr. Stephen Cambone served as Under Secretary of Defense for Intelligence from 2003 to 2007. Prior to this assignment, Cambone had served as staff director secretary for two high-level commissions chaired by Donald Rumsfeld, the 1998 Commission to Investigate the Ballistic Missile Threat to the United States and the 2000–2001 Commission to Assess United States National Security Space Management and Organization (Space Commission) that remains the most important and influential examination of national security space (NSS) issues ever undertaken. After he became Secretary of Defense in 2001, Rumsfeld made Cambone his "go to" person for space, regardless of the office Cambone held, and eventually placed him in the Under Secretary of Defense for Intelligence position created in March 2003. This move was inconsistent with a major recommendation from the Space Commission Rumsfeld chaired before becoming Secretary that called for creation of an Under Secretary of Defense for Space, Information, and Intelligence. Rumsfeld's move did not help to institutionalize centralized authority and responsibility for NSS within the Office of the Secretary of Defense (OSD), undermined the Space Commission's vision for organization and management of NSS, and contributes to continuing friction between OSD branches as well as overlaps, gaps, and unclear lines of authority and responsibility between OSD and the DoD Executive Agent (EA) for Space.

James E. Carter Jr. (October 1, 1924–)

Jimmy Carter was President of the United States from 1977 to 1981. Early in his administration, Carter publicly announced the United States had proposed ASAT negotiations to the Soviets and at the same time secretly issued Policy Review Memorandum (PRM)-23 which directed the NSC Policy Review Committee (PRC) to review existing policy. The Soviets soundly rejected the administration's comprehensive proposal but did agree to set up various working groups to discuss specific arms-control issues, including ASAT issues. Carter's two-track policy for simultaneously pursuing ASAT development and ASAT arms control was in place by fall 1977. U.S. and Soviet negotiators met for three rounds of ASAT talks: June 8–16, 1978, in Helsinki; January 23–February 16, 1979; in Bern, and April 23–June 17, 1979, in Vienna; but failed to come close to reaching any agreement, and the negotiations were ended after increasing tensions following the Soviet invasion of Afghanistan. During 1978, the Carter Administration developed two comprehensive, classified space-policy statements: Presidential Directives (PDs)-37 and 42 were motivated by top-down national security considerations and driven by the cumulative impact of incremental technological improvements in space systems, concerns with ASAT capabilities, satellite survivability, and the military potential of the STS. PD-37, "National Space Policy," was signed by Carter on May 11, 1978; the fact sheet stated that the directive: encouraged lowering space classification levels where possible, initiated a program to increase survivability for military space systems, and detailed the two-track U.S. approach to ASAT. On October 1, 1978, President Carter publicly discussed the importance of spysats in verifying arms-control agreements, marking the first official break with the blackout policy promulgated in 1962. National Security Adviser Zbigniew Brzezinski signed PD-42, "Civil and Further National Space Policy," on October 10, 1978, and a fact sheet was released the next day. The fact sheet indicated three components for U.S. civil space policy: (1) evolutionary space activities unique to or more efficiently accomplished in space; (2) a balanced strategy of applications, science, and technology development; and (3) emphasis that it was neither feasible nor necessary to initiate a challenging space program comparable to Apollo at this time.

William J. Clinton (August 19, 1946–)

Bill Clinton served as President of the United States from 1993 to 2001. The National Space Council was disbanded early in the Clinton presidency, but the administration remained very active in national space policy, using the NSC and the National Science and Technology Council (NSTC) as the lead organizations for developing space policy. Clinton issued policy on Foreign Access to Remote Sensing Space Capabilities in Presidential Decision Directive in (PDD)-23 in March 1994. PDD-23 was designed to encourage U.S. worldwide preeminence in commercial imagery while protecting national security by allowing sales of high-resolution imagery or even entire imagery systems in tightly controlled cases. He issued Convergence of U.S.-Polar Orbiting Environmental Satellite Systems in NSTC-2 in May 1994. NSTC-2 directed DoD, the Department of Commerce's National Oceanic and Atmospheric Agency (NOAA), and NASA to work together on a single environmental monitoring system that would meet their requirements and save money. The National Polar-Orbiting Environmental Satellite System (NPOESS) program resulted from this directive; NPOESS repeatedly ran into problems in meeting its requirements on time and on budget resulting in the program being ended in 2010. Clinton signed Landsat Remote Sensing Strategy in NSTC-3 in May 1994 that reflected the failure of Landsat 6 in 1993, transitioned Landsat 7 from DoD to NASA, and established a plan for the U.S. government to continue operating Landsat satellites and maintaining an archive of Landsat-type data. And he released a National Space Transportation Policy in NSTC-4 in August 1994, an attempt to sustain and revitalize U.S. launch systems by making DoD the lead agency for improving the ELV fleet and NASA the lead agency for developing next generation reusable launch vehicles; the Evolved ELV (EELV) and the X-33 and X-34 programs resulted from this policy. In addition, the policy made the Departments of Transportation and Commerce responsible for promoting innovative arrangements to encourage a viable U.S. space-transportation industry and stipulated that the U.S. government would purchase U.S. launch products and services and not launch government payloads on foreign vehicles except after being granted an exemption by the President. On March 29, 1996, the Clinton Administration issued a new GPS policy indicating the United States would continue to provide GPS service without direct user fees, encourage worldwide acceptance and integration of GPS, and discontinue deliberately degrading the accuracy of signals provided to nonmilitary users (Selective Availability or SA) within a decade; Clinton discontinued SA in May 2000. Finally, on September 19, 1996, the White House released a fact sheet about PDD-49, "National Space Policy," that reemphasized the United States would maintain a leadership role in space by supporting a strong, stable, and balanced program across all space-activity sectors. Clinton's policy built from the component space policies discussed above and was once again very consistent with previous national space policies stretching back to the Eisenhower Administration. Areas of different or increased emphasis included responsibilities for: NASA and other agencies to develop sensors and acquire data to closely observe and make predictions about Earth's environment; DoD to foster integration and interoperability of satellite control for all governmental space activities; and the Arms Control and Disarmament Agency to identify arms-control opportunities for equitable and effectively verifiable measures that would enhance the security of the United States and its allies.

Hugh L. Dryden
(July 2, 1898–December 2, 1965)

Dr. Hugh Dryden was director of the National Advisory Council on Aeronautics (NACA) from 1947 to 1958 and held high-level positions in NASA from 1958 to 1965. Dryden had strongly urged President Eisenhower to make NASA, rather than DoD, primarily responsible for the manned spaceflight. In July 1958, Dryden met with Advanced Research Projects Agency (ARPA) Director Roy Johnson and Secretary of Defense Neil McElroy to discuss the future management of manned space programs, but they were unable to resolve their organizational differences. Eisenhower formally gave NASA primary authority over U.S. manned spaceflight efforts in August 1958 and by November this effort had evolved into Project Mercury.

Dwight D. Eisenhower
(October 14, 1890–March 28, 1969)

Dwight Eisenhower served as President of the United States from 1953 to 1961. Eisenhower's perceptions of space and strategic issues were strongly influenced by the top secret Technological Capabilities Panel (TCP) he commissioned in March 1954, choosing James Killian, President of the Massachusetts Institute of Technology (MIT), as its chairman. TCP recommendations and the desire to reconcile conflicting desires and programs led to America's first space policy, NSC-5520, a secret document Eisenhower signed on May 27, 1955. Contrary to the public impression created in the crisis atmosphere sparked by the shock of the Soviet *Sputnik* I triumph, under NSC-5520 the Eisenhower Administration laid out a logical and comprehensive approach designed to advance U.S. national interests at the opening of the space age. Eisenhower's primary space-policy goal was to establish freedom of space by creating an international legal regime that allowed satellite overflight in order to open up the closed Soviet state via satellite reconnaissance. The Eisenhower Administration laid the foundation for two separate U.S. space programs: a hidden top-priority effort to develop spy satellites, and an open effort emphasizing use of space for peaceful purposes such as space science and international cooperation. Of course, a secret policy was of little use in defusing the crisis in public confidence that developed after the *Sputniks* and Eisenhower pursued public policies to highlight his concern with the situation, including appointment of Killian as the first President's Science Advisor and creation of ARPA. The Eisenhower Administration also pushed development of ICBMs: Eisenhower gave the Atlas the highest national priority for R&D in September 1955 and thereafter the ICBM consistently was developed about as fast as technologically possible; it was first test flown successfully in November 1958 and achieved initial operational capability (IOC) in September 1959.

Howell M. Estes III
(December 16, 1941–)

Howell Estes was commander of United States Space Command (CINCUSSPACE) from 1996 to 1998. Major initiatives as

CINCUSSPACE included an unsuccessful push to have space designated as USSPACECOM's area of responsibility (AOR), commissioning a study on spacepower theory, and in 1998 publishing *Long Range Plan: Implementing USSPACECOM Vision for 2020*. After retirement, Estes served as a member of the Space Commission that delivered a very important and influential report on January 11, 2001.

James C. Fletcher
(June 5, 1919–December 22, 1991)

Dr. James Fletcher served as Administrator of NASA from 1971 to 1977 and 1986 to 1989. On January 5, 1972, NASA Administrator Fletcher went to the Western White House to brief President Nixon about the shuttle design and to be present when the decision to approve the STS was publicly announced. He also was chairman of the 1983–1984 Defensive Technologies Study that examined emerging missile-defense technologies and helped provide support for creating the Strategic Defense Initiative Office in 1984.

Gerald R. Ford
(July 14, 1913–December 26, 2006)

Gerald Ford was President of the United States from 1974 to 1977. A 1976 study for the NSC concluded that a U.S. ASAT would not enhance the survivability of U.S. satellites by deterring use of the Soviet ASAT because the United States was more dependent on space than the Soviets. The report also concluded, however, that a U.S. ASAT could be used to counter the threat to U.S. forces posed by Soviet space-based targeting systems such as Radar Ocean Reconnaissance Satellites (RORSATs) and Electronic Intelligence Ocean Reconnaissance Satellites (EORSATs) and that the development of a U.S. ASAT could serve as a "bargaining chip" in possible U.S.-U.S.S.R. ASAT arms-control negotiations. Ford issued National Security Decision Memorandum (NSDM)-333 in the fall of 1976; it directed DoD to work harder on satellite survivability and resulted in the creation of a separate Systems Program Office (SPO) for Space Defense Programs at the Air Force's Space and Missile Systems Organization (SAMSO) and increased funding for these types of efforts. In one of the final acts of his presidency, on

January 18, 1977, Ford approved NSDM-345, directing DoD to develop an operational ASAT system. This initiated the air-launched miniature homing vehicle (MHV) ASAT program and set the stage for two-track ASAT negotiations during the Carter Administration.

T. Keith Glennan
(September 8, 1905–April 11, 1995)

Keith Glennan was the first Administrator of NASA, serving from 1958 to 1961. In October 1958, Glennan worked out a deal to transfer the Army-sponsored Jet Propulsion Laboratory (JPL) at the California Institute of Technology and the von Braun team at the Army Ballistic Missile Agency (ABMA) to NASA. Met with Army resistance, Glennan and the National Aeronautics and Space Council (NASC) worked out a compromise in December that made JPL a part of NASA and allowed the von Braun team to remain under ABMA but work on the Saturn booster under contract to NASA. Glennan signed a NASA-DoD Memorandum of Understanding (MOU) on Project Mercury on November 20, 1958. According to the terms of the MOU, DoD was to cooperate with NASA on the conduct of the program, NASA was to use the resources of DoD, and ARPA was to contribute $8 million in fiscal year (FY) 1959 funds to NASA.

Daniel S. Goldin (July 23, 1940–)

Daniel Goldin holds the current record for longest tenure as NASA Administrator, serving from June 1992 to January 2001. Goldin's "faster, better, cheaper" approach helped transform NASA during a period of reduced funding and lower enthusiasm for large space programs. He was also instrumental in bringing the Russian Federation aboard the space station in September 1993 and transforming this effort into the current International Space Station.

Daniel O. Graham
(April 13, 1925–December 31, 1995)

Daniel Graham was a Lieutenant General in the United States Army widely credited with encouraging movement toward SDI.

Graham's thinking on American strategy moved away from the mutual assured destruction (MAD) paradigm and focused on the potential of space. After retiring from the Army, Graham founded High Frontier, a political action organization, in 1981, and published *High Frontier: A New National Strategy* in March 1982. Later he focused on development of single-stage-to-orbit boosters such as the DC-XA Delta Clipper test vehicle that was renamed Clipper Graham in 1996 in his memory.

Michael D. Griffin (November 1, 1949–)

Dr. Michael Griffin was Administrator of NASA from 2005 to 2009. While at the Johns Hopkins University Applied Physics Laboratory (APL) in the 1980s, Griffin helped design the successful Delta 180 experiment on missile-defense technologies for the Strategic Defense Initiative Office (SDIO). After leaving APL in 1986, he served as the SDIO deputy for technology, then as the chief engineer and later associate administrator for exploration at NASA Headquarters. As NASA Administrator, Griffin developed the Constellation program to pursue President Bush's January 2004 "Vision for Space Exploration." The Constellation program sought to advance U.S. scientific, security, and economic interests through a robust space-exploration program including a sustained and affordable human and robotic program to explore the solar system and beyond, starting with a human return to the Moon by the year 2020, in preparation for human exploration of Mars and other destinations. However, the Constellation program was not funded in the NASA budget the Obama Administration submitted to Congress in February 2010.

Richard C. Henry (1925–)

Lieutenant General Richard Henry was Air Force System Command Space Division Commander and became Commander of the Space and Missile Systems Organization (SAMSO) in 1978. Writing as a Major in 1961, Henry helped to inject some restraint and perspective on the Air Force role in space. He noted that the aerospace concept needed to be approached similarly to the Air

Force's approach to airpower and developed what is probably the first open exposition of a space doctrine derived from the environmental characteristics of space. Henry summarized his environmental doctrine by providing a comparison of the relative advantages and disadvantages for orbital systems, indicating that manned recoverable spacecraft and a space station were near-term military requirements in space. In June 1982, Henry emphasized that all space systems served a national purpose and not just military missions, argued against the need for a separate space command, and was concerned that the commercial sector would outweigh the military sector. Henry also identified three central issues related to maintaining assured access to data streams from space for force enhancement: space system survivability, hardware for connectivity with major military systems, and the direct command and control links between headquarters and unit level forces possible through space-based communications relays. Significantly, Henry focused on space-based force enhancement capabilities almost exclusively rather than even discussing the potential for space control or using space as the high ground.

Robert T. Herres
(December 1, 1932–July 24, 2008)

Air Force General Robert Herres served as the first Commander of United States Space Command from 1985 to 1987 and as Vice Chairman of the Joint Chiefs of Staff from 1987 to 1990. While Commander of United States Space Command (CINCUSSPACE), Herres opposed creation of a fourth military department, a Department of Space, within DoD. Explaining USSPACECOM's missions, Herres emphasized that even if there were a Department of Space, the forces of this new department would still be employed through USSPACECOM. He also provided some support for the MHV ASAT program by emphasizing the capabilities of the Soviet ASAT system and the threat posed to the Navy by Soviet RORSAT and EORSAT targeting systems. Herres left USSPACE-COM on February 6, 1987, to become the first Vice Chairman of the JCS, a position created by the 1986 Goldwater-Nichols DoD reorganization act.

Robert Jastrow
(September 7, 1925–February 8, 2008)

Dr. Robert Jastrow was a theoretical physicist and cosmologist who joined NASA when it was established in 1958, served as first chairman of the Lunar Exploration Committee, chief of the Theoretical Division, and became the founding director of NASA's Goddard Institute for Space Studies at Columbia University in 1961, serving until his retirement from NASA in 1981. Later, Jastrow became a strong supporter of strategic defenses, publishing *How to Make Nuclear Weapons Obsolete* in 1983 and helping found the George C. Marshall Institute in 1984.

Lyndon B. Johnson
(August 27, 1908–January 22, 1973)

Lyndon Johnson served as a senator from Texas from 1949 to 1961 and was majority leader from 1955 to 1961. He was then Vice President of the United States from 1961 to 1963 and President from 1963 to 1969. Johnson used his position on the Senate Armed Services Committee to convene hearings that ran from November 1957 to January 1958 and became one of the most important factors in shaping American attitudes on space and security issues in the period immediately after the *Sputniks*, events tailor-made for him to challenge President Eisenhower and catapult him into national prominence as he prepared his bid for the 1960 presidential election. Johnson's hearings can also be seen as marking the end of the initial *Sputniks* crisis period because they were the focal point in moving America away from the calm and restrained approach favored by Eisenhower toward a far more active and concerned response; Johnson emphasized that space was the ultimate position that must be held at all costs in a well-publicized speech to the Democratic caucus on January 7, 1958. Following Yuri Gagarin's triumphant orbital flight in April 1961, President Kennedy turned to Vice President Johnson to develop the most appropriate response options. Johnson quickly centered on a Moon race as the best response to the Soviets and his recommendation led directly to Kennedy's May 25, 1961, Moon landing challenge. As

President, Johnson publicly revealed the existence of a U.S. ASAT capability on September 17, 1964. At a press conference on August 25. 1965, President Johnson formally approved the development of the Manned Orbiting Laboratory (MOL). Toward the end of his administration, Johnson supported the push at the UN that would culminate in the 1967 Outer Space Treaty.

John F. Kennedy
(May 29, 1917–November 22, 1963)

John Kennedy served as a senator from Massachusetts and President of the United States from 1961 to 1963. President Kennedy started formulating his space policy before entering office, asking Dr. Jerome B. Wiesner of MIT (who would become his Science Adviser) to head the Ad Hoc Committee on Space and tasking this group to study the structure for and the direction of U.S. space efforts. Kennedy met with this group on January 10, 1961, and an unclassified version of their report was released the next day that recommended revitalizing the NASC, putting primary emphasis on space-science missions, and strongly warned against attempting to race the Soviets for manned space spectaculars. Following the Soviet success with the first orbital spaceflight, however, Kennedy felt compelled to show America's strength by beating the Soviets in some race to demonstrate U.S. science and technology prowess. Kennedy took advice from Vice President Johnson and the NASC and decided on a prestige-based Moon race as the best way to attempt to beat the Soviets in a major space spectacular, issuing his Moon-landing challenge on May 25. He also advanced three other major initiatives in space policy: the space-law development process that would culminate in the 1967 Outer Space Treaty (OST), secret efforts to protect and further legitimize the emerging spy-satellite regime, and a quest for arms control in space. The administration took a two-track approach to ASAT development and arms-control efforts by deploying a minimum number of ASATs to mitigate against a Soviet orbital nuclear-weapon threat while simultaneously pursuing arms-control efforts to ban such weapons in space and thereby remove a major incentive for deploying ASATs. Early in the Kennedy Administration, however, efforts to achieve space arms control were severely hampered by a lack of interagency coordination. Accordingly, on May 26, 1962, Kennedy issued National Security Action Memorandum (NSAM) 156—an implicit critique of the NASC and a request for the Department of State to create a high-level coordinating body (known as the NASM 156 Committee) to address this problem. The primary responsibility of the NASM 156 Committee was to develop policies designed to protect and legitimize U.S. spysats, but this group was also chiefly responsible for creating the U.S. initiatives aimed at banning nuclear weapons from outer space.

George A. Keyworth II
(November 30, 1939–)

Dr. George Keyworth served as Science Adviser under President Reagan from 1981 to 1985. He chaired the Senior Interagency Group or SIG (Space) at the White House that was often the scene of intense debates over the proper focus of future U.S. space efforts. Generally, Keyworth remained unconvinced of the strategic utility of the large-scale, permanently manned space station proposed by NASA Administrator Beggs. Keyworth and his supporters felt that NASA should first concentrate on getting the STS to achieve its performance requirements before diverting attention to the next space spectacular. He was a strong supporter of SDIO and technical efforts to develop ballistic missile defenses.

James R. Killian Jr.
(July 24, 1904–January 29, 1988)

Dr. James Killian was President of the Massachusetts Institute of Technology (MIT) before accepting President Eisenhower's invitation to chair the top secret Technological Capabilities Panel (TCP) in March 1954 and later served as the nation's first Special Adviser to the President for Science and Technology (Science Adviser) between November 1957 and 1959. Part of Killian's work on the TCP culminated in a secret November 1954 Oval Office meeting where President Eisenhower authorized initiation of a high-technology strategic reconnaissance program including a high-altitude jet glider then on the drawing board at Clarence "Kelly" Johnson's Lockheed Skunk Works that became the U-2 and development of reconnaissance satellites that because the Air Force's Weapons System (WS)-117L program and the CIA's Corona program. Killian was concerned that military considerations would dominate space and strongly urged Eisenhower to create a civilian space agency that would focus on space science and human spaceflight. As Eisenhower's top space-policy adviser, Killian's views were very influential; in early 1958 he supervised the task of writing the proposed legislation to create a civilian space agency that became NASA. In dividing military and civil space responsibilities, Killian favored having NASA design and build the capsules for manned spaceflight and ARPA concentrate on the boosters required for this mission. Killian's perceptions on the relationships between space and national security were also clearly illustrated during the deliberations over the comprehensive national space policy in NSC 5814/1. Killian ran studies by an Ad Hoc Subcommittee on Outer Space forming the basis for the Planning Board's secret draft of NSC 5814 completed on June 20, 1958. The final NSC-level debate on NSC 5814 came at the NSC meeting on August 14, where the primary discussion centered on the level of priority that should be given to military space missions; Killian successfully argued to delete a paragraph from NSC 5814, removing a clear statement that military space activities would receive priority. Finally, Killian strongly urged Eisenhower to make NASA primarily responsible for the manned mission and this approach resonated with Eisenhower's concerns with space for peaceful purposes and his desire to avoid duplication of effort and the costly space race proposed by the Air Force.

George Kistiakowsky (November 18, 1900–December 7, 1982)

Dr. George Kistiakowsky worked on the Manhattan Project during World War II and became President Eisenhower's Special Assistant for Science and Technology in 1959, serving in this position until Eisenhower left office in January 1961. Kistiakowsky believed that the Air Force was putting too much effort into the electro-optical data return Samos reconnaissance satellite program, which was based on technologies he felt would not mature for some time, and disrupting the entire spysat development effort. In

May 1960, Eisenhower directed Kistiakowsky to study what corrective actions might be necessary. Secretary of Defense Thomas Gates and Kistiakowsky decided on a study committee composed of three people: Under Secretary of the Air Force Joseph Charyk, Deputy Director of Defense Research and Engineering (DDR&E) John Rubel, and Kistiakowsky. This Samos Panel reported their recommendations to the NSC on August 25. Their primary recommendation, immediate creation of an organization to provide a direct chain of command from the Secretary of the Air Force to the officers in charge of each spysat project, was enthusiastically supported by Eisenhower and led to the formal establishment in August 1961 of the highly classified National Reconnaissance Office (NRO). Later, Kistiakowsky staunchly opposed the Air Force's SAINT ASAT proposal but supported a plan to cancel the Saturn C-1 booster, transfer the von Braun team from ABMA to NASA, and transfer authority for all military booster development to the Air Force, allowing the military to develop its own technologies.

Hans M. Mark (June 17, 1929–)

In July 1977, Dr. Hans Mark, who had been Director of NASA's Ames Research Center, became Under Secretary of the Air Force and NRO Director; during the Reagan Administration he served as NASA Deputy Administrator from 1981 to 1984. As an avid human spaceflight enthusiast who believed the STS was an essential step toward future space stations and exploration, Mark attempted to push the STS and a greater space emphasis throughout the Air Force and was instrumental in lining up DoD support for the STS when it was in danger of being canceled by the Carter Administration. Mark's efforts and the military potential of the STS revived Air Force interest in space and in possible military-man-in-space applications. In April 1981, as a result of a January 1980 challenge issued by then-Secretary of the Air Force Mark to the USAF Academy to determine a military doctrine for space, the Academy hosted what is probably the most important conference devoted to military space doctrine during the Cold War. The symposium included a keynote address by Mark and attracted a large audience of military space experts from around the Air Force, the other services, and civilian space organizations. Several important space-doctrine issues were addressed at the symposium and they helped reinforce support for creating a new space command within the Air Force. Early in the Reagan Administration, when Mark was NASA Deputy Administrator, the Air Force was interested in producing another of the lighter weight and more capable orbiters but was unwilling to use DoD funds for a fifth orbiter. Meanwhile, NASA was less supportive of the need for a fifth orbiter, largely because Mark had privately agreed with Administrator Beggs that NASA should push a permanently manned space station as the nation's new major civil space goal and was therefore unwilling to take on other major new projects at this time. Thus, according to Mark, NASA lost out in the debates which culminated in National Security Decision Directive (NSDD)-42 in two important ways: first, NASA was unable at this time to gain an administration commitment to a space station as the nation's next major space goal; and second, the language of the directive also opened the door through which the NRO would eventually push spy satellites off the STS and onto the CELV.

Neil H. McElroy (October 30, 1904–November 30, 1972)

Neil McElroy served as the sixth Secretary of Defense from October 1957 to December 1959. Taking office on October 9, 1957, McElroy's term was largely shaped by the *Sputniks*, efforts to speed U.S. space capabilities, the missile gap, and work to reorganize DoD. McElroy accelerated Air Force Thor and Army Jupiter IRBM development and sped up development of the Navy solid-fuel Polaris IRBM and the Air Force liquid-fuel Atlas and Titan ICBMs. In February 1958, he authorized the Air Force to begin development of the Minuteman, a solid-fuel ICBM to be deployed in hardened underground silos and become operational in the early 1960s. McElroy centralized control of space R&D efforts within the Advanced Research Project Agency (ARPA) created on November 15, 1957, initially charging the new agency to take control of all military space programs and move ahead quickly to avoid the situation the United States faced after *Sputnik*. He attempted to use ARPA to consolidate and focus military space efforts and rejected a proposal from the JCS for unified space command on September 18, 1959. On January 16, 1958, McElroy assigned the Army primary responsibility for developing an ABM system over

Air Force objections but the Army's program, known as Nike-X, made only halting progress in the early 1960s. McElroy considered creation of a strong Director of Defense Research and Engineering (DDR&E) to be one of the most important aspects of the Defense Reorganization Act of 1958.

Robert S. McNamara
(June 9, 1919–July 6, 2009)

Robert McNamara was Secretary of Defense from January 1961 to February 1968. McNamara sought efficiencies and consolidations in many areas including space; he attempted to centralize military space efforts through DoD Directive 5160.32, "Development of Space Systems," issued on March 6, 1961. He wanted to exert more direct and tighter control over DoD space efforts and believed centralizing DoD space R&D within the Air Force to be the easiest way to pursue this objective. McNamara made a secret decision in May 1962 directing the Army to develop modified Nike Zeus interceptors as an ASAT system (Program 505) and in November 1962 ordered the Air Force to modify Thor IRBMs as an additional ASAT capability (Program 437); both these systems used nuclear warheads. Following the initiation of Project Apollo, McNamara and other leaders within OSD understood that NASA's Moon race was the highest U.S. space priority and took a cautious "building block" approach that reflected their skepticism about the rationale and utility of Air Force efforts to build a large manned military space presence. Attempting to gain a greater role or control by the Air Force over Project Gemini, McNamara was able to sign a January 1963 agreement with NASA to allow DoD experiments on certain "Blue Gemini" missions. Finally, he cancelled the Air Force's X-20 Dynamic Soaring (Dyna-Soar) program on December 10, 1963, announcing at the same time that the service would be responsible for developing the Manned Orbiting Laboratory (MOL).

John B. Medaris
(May 12, 1902–July 11, 1990)

Army Major General John B. Medaris became the first commander of the Army Ballistic Missile Agency (ABMA) in February 1956.

ABMA included the team of German engineers led by Wernher von Braun that had developed the V-2, the world's first ballistic missile; their efforts in America to advance missile technology and launch satellites were often stymied by a range of bureaucratic and political factors. Following the success of *Sputnik* I, Medaris asserted that ABMA's Project Orbiter could place a satellite in orbit within 90 days and speeded preparations to attempt to launch a satellite as much as possible following the spectacular failure of America's first satellite launch attempt with the December 6 launch-pad explosion of Vanguard TV-3. The first U.S. satellite, *Explorer* I, was launched on January 31, 1958, aboard a Jupiter-C (Juno) launch vehicle developed by ABMA. Following this success, Medaris believed ABMA would be rewarded and remain under Army control but this did not align with President Eisenhower's space-for-peaceful-purposes approach or the widespread support for creating a strong civilian space agency. For the remainder of his Army career, Medaris waged a vigorous but ultimately unsuccessful campaign to keep the von Braun team. Medaris retired in February 1960 and President Eisenhower presided over the transfer of the von Braun team and many ABMA facilities to NASA's Marshall Space Flight Center on July 1, 1960. After retirement, Medaris published *Countdown for Decision*, a bitter attack on the Eisenhower Administration policies that had held back and then dismantled ABMA.

Thomas S. Moorman Jr. (1940–)

General Tom Moorman attained the highest position of any space general thus far by serving as Vice Chief of Staff of the Air Force from July 1994 to August 1997. Prior to this position, Moorman served as Vice Commander and Commander of Air Force Space Command. In 1993–1994 Moorman headed a launch vehicle modernization study that recommended Air Force and NASA launch programs be separated and the Air Force develop a new or evolved expendable launch vehicle (EELV); this approach was formalized by the Clinton Administration's National Space Transportation Policy (NSTC-4) in August 1994. After retiring from the Air Force, Moorman served as a member of the 2000–2001 Space Commission, the most important and influential study on NSS management and organization yet undertaken.

Richard M. Nixon
(January 9, 1913–April 22, 1994)

Richard Nixon served as Vice President during the Eisenhower Administration and was President from January 1969 to August 1974. Since he was not a particularly strong supporter of NASA or the Apollo program, it is somewhat ironic that all six U.S. Moon landings took place during his tenure. Shortly after entering office, Nixon established a Space Task Group (STG) to complete a comprehensive review of future plans for the U.S. space program; the STG was comprised of Vice President Spiro Agnew, Acting NASA Administrator Thomas Paine, Secretary of Defense Melvin Laird, and Science Adviser Lee A. DuBridge. On September 15, 1969, the STG presented Nixon with options for post-Apollo U.S. civil space plans including a manned mission to Mars by 1985, a 50-man space station in orbit around Earth, a smaller space station in orbit around the Moon, a lunar base, a space shuttle to service the Earth space station, and a space tug to service the lunar stations. By March 1970, Nixon endorsed only development of a shuttle, leaving a space station or a Mars mission contingent on successful completion of a shuttle program. After very intense scrutiny from OMB during 1971, Nixon approved a space shuttle program with a thrust-assisted design on January 5, 1972. The Nixon Administration initially supported the Manned Orbiting Laboratory (MOL), requesting $525 million for the program in FY 1970, but this support soon evaporated in the face of the Vietnam War and other DoD requirements; Nixon officially canceled the MOL on June 10, 1969. The decision to terminate MOL was made at a White House meeting of Office of Management and Budget (OMB) representative Robert Mayo, National Security Adviser Henry A. Kissinger, and President Nixon; however, as they made clear in subsequent congressional testimony, Secretary of Defense Melvin Laird and the JCS were not consulted prior to this decision.

Barack H. Obama (August 4, 1961–)

Barack Obama is the 44th and current President of the United States. During his presidential campaign he pledged to support NASA, not weaponize space, and reestablish a space council at the White House. In the budget submitted to Congress in February 2010 the

President did not request funding for NASA's Constellation program to go forward as originally planned; his budget would also terminate the National Polar-Orbiting Operational Environmental Satellite System (NPOESS) by splitting it into separate DoD and NOAA missions.

Thomas O. Paine
(November 9, 1921–May 4, 1992)

Dr. Thomas Paine served as NASA Deputy Administrator, Acting Administrator, and Administrator between January 1968 and September 1970. During 1969, he served on the Space Task Group (STG) that completed a comprehensive review of future options for the U.S. space program. As Administrator, Paine emphasized development of a large, highly capable space shuttle. In 1985 the White House chose Paine to chair the National Commission on Space that published *Pioneering the Space Frontier* in May 1986, a report strongly advocating highly ambitious plans for space exploration and settlements.

Samuel C. Phillips (February 19,
1921– January 31, 1990)

Samuel Phillips was an Air Force General who served as Director of NASA's Apollo Lunar Landing program from 1964–69, Commander of the Space and Missile Systems Organization (SAMSO), Director of the National Security Agency, and Commander of Air Force Systems Command. In the Apollo program, Phillips was famous for his hands-on style of leadership, nonstop on-site contractor inspections, and comprehensive grasp of all program details. Wernher von Braun praised Phillips as the person most responsible for making all the many parts of Apollo work together on time to accomplish the mission.

John L. Piotrowski (1934–)

John Piotrowski was an Air Force General who served as CINCSPACE from 1987 to 1990. He used this position to provide what is probably the strongest space advocacy from any military member during the Cold War. Piotrowski supported development of force enhancement systems such as a space-based radar (SBR) but saved his greatest enthusiasm for high-ground concepts such as SDI and for programs to support the space-control school including the MHV ASAT system, improved space-surveillance capabilities, and measures to enhance U.S. satellite survivability. Piotrowski was even more forceful and persistent in describing the comprehensive nature of the Soviet space threat and arguing the need for a robust U.S. space-control infrastructure. During congressional testimony in 1987, he highlighted significant Soviet advantages in almost all categories of military space activities and emphasized that an operational ASAT capability was required for space control.

Donald A. Quarles
(July 30, 1894–May 8, 1959)

Donald Quarles held a succession of increasingly important positions in the Air Force and OSD, including Assistant Secretary of Defense for Research & Development beginning in September 1953, Secretary of the Air Force in August 1955, and Deputy Secretary of Defense in May 1957. Quarles was one of the few officials who understood how complex and conflicting U.S. policy and technology development goals were before the opening of the space age; he attempted to harmonize these factors as primary author of America's first space policy, NSC-5520, signed in May 1955. On October 8, 1957, four days after launch of *Sputnik* I, Quarles requested that DoD study the possibility of using the Jupiter-C IRBM as a quicker and surer way to launch America's first satellite. Once processes were established to legitimize reconnaissance satellite and develop spysats, Quarles turned his attention to management and organizational structures. In October 1958, he worked out a deal to transfer the Army-sponsored Jet Propulsion Laboratory (JPL) at the California Institute of Technology and the von Braun team at ABMA to NASA. During 1959, Quarles became increasingly convinced that the Saturn program and the von Braun team did not belong in the Army.

Ronald W. Reagan
(February 6, 1911–June 5, 2004)

Ronald Reagan served as President of the United States from 1981 to 1989; his administration devoted a great deal of attention to space issues and generated more official space-policy statements than any other during the Cold War. Reagan's space-policy statements covered a wide range of topics but generally emphasized military space potential and space commercialization efforts to a greater degree than any previous administration; they also represent another instance where civilian leadership rather than the military led the push for greater investigation of the military potential of space. On November 13, 1981, Reagan signed National Security Decision Directive (NSDD)-8, reaffirming the space-transportation policies of the Ford and Carter Administrations by stating that the STS would be the primary launch system. His administration next developed NSDD-42, a comprehensive space policy publicly announced on July 4, 1982, and this policy served as foundational space-policy guidance for the bulk of his two terms in office. Reagan's Star Wars challenge of March 23, 1983, changed the terms of the strategic debate within the United States and worldwide by reintroducing a fundamental strategic concept while simultaneously outflanking the nuclear freeze movement; it also meant military space issues were again as conceptually wide open as they had been in October 1957. In 1983–1984, the administration completed three important studies on SDI and set in motion the bureaucratic structure for SDIO. The Reagan Administration did not initially value the arms-control process as highly as had its predecessors and it did not link spysats as directly with national technical means of verification; yet it was the first to use the products from spysats (and other sources) systematically and publicly to emphasize the extent of the continuing Soviet military expansion as a part of its broader domestic and international efforts to gain support for the U.S. defense buildup. The Reagan Administration also was required to mount significant efforts to counter congressional restrictions on testing the MHV ASAT system and beginning in 1986 initiated a major joint DoD-NASA space launch technology development effort known as the X-30 or National Aerospace Plane (NASP).

Donald H. Rumsfeld (January 9, 1932–)

Donald Rumsfeld was the 21st Secretary of Defense from January 2001 to December 2006 and had previous service as 13th Secretary of Defense, White House Chief of Staff, U.S. Ambassador to NATO, and U.S. congressman. In 2000, he served as chairman of the Commission to Assess United States National Security Space Management and Organization (Space Commission), resigning this position prior to the release of the report on January 11, 2001, because he had been nominated to become Secretary of Defense. Due to a sweeping charter, powerful members, and comprehensive recommendations, the Space Commission is the most important and influential NSS review yet undertaken. It is ironic, however, that Rumsfeld as Secretary of Defense devoted almost no time to space issues and did so little to implement the Space Commission's recommendations. The Air Force and most other NSS stakeholders moved fairly quickly and effectively to implement Space Commission recommendations but other organizations did not and there was much less focus on NSS issues from the DoD and IC following the 9/11 attacks. Air Force implementation actions included making the Commander of Air Force Space Command (AFSPC) a four star billet that need not be flight rated and moving AFSPC out from the combatant command authority of USSPACECOM; designating the Under Secretary of the Air Force as the Director of the NRO, Air Force Acquisition Executive for Space, and DoD Executive Agent (EA) for Space; aligning the Space and Missile Systems Center (SMC) underneath AFSPC instead of Air Force Materiel Command; and establishing a major force program (MFP) accounting category for the NSS budget. Rumsfeld authorized a very significant change to the Space Commission's vision for NSS management and organization on October 1, 2002, when USSPACECOM was merged into U.S. Strategic Command (USSTRATCOM); although originally justified by the desire to create U.S. Northern Command after the 9/11 attacks and initially described as a joining of equals, in practice this major organizational shift quickly amounted to the absorption of USSPACECOM into USSTRATCOM and left very few vestiges of the original USSPACECOM. Rumsfeld next made Dr. Stephen Cambone his "go to" person for space, regardless of the office Cambone held, and eventually placed him in the Under Secretary of Defense for Intelligence position created in March

2003. Finally, in July 2005, Rumsfeld decided that incoming Under Secretary of the Air Force and DoD EA, Dr. Ronald Sega, would not, as had his predecessor, also be Director of the NRO; this changed the organization structure of NRO that had existed since 1961, undermined a major Space Commission recommendation for better DoD-Intelligence Community collaboration, and again called into question the strength of Rumsfeld's commitment to the recommendations from the commission he chaired.

Bernard A. Schriever (September 14, 1910–June 20, 2005)

Bernie Schriever was the Air Force's point man for developing missile and satellite capabilities at the beginning of the space age, work that had to be carefully structured to comply with President Eisenhower's space for peaceful purposes policy and that often brought him into conflict with supporters of manned bombers such as General Curtis LeMay. In August 1954, Schriever took command of the new Western Development Division (WDD) that had been created to expedite development of the Atlas ICBM; at WDD Schriever initiated the WS-117L satellite program and also pioneered new methods of systems management such as moving beyond the traditional Air Force contractor model and exploring concurrent development processes. After Sputnik, Schriever redoubled his efforts to develop the Atlas and WS-117L. He opposed creation of a new space agency by arguing that the Air Force already had the required technology and proposing in January 1958 a comprehensive program for investigating the feasibility of manned spaceflight and launching lunar probes beginning in 1959. Schriever was also openly and consistently critical of ARPA; in congressional testimony in April 1959 he recommended ARPA be abolished. In October 1960, Schriever asked Trevor Gardner to chair the Air Force Space Study Committee; this group delivered a secret report in March 1961 providing a ringing endorsement of the high-ground and space-control schools and ignoring NASA by calling for the new Air Force Systems Command (AFSC) to spearhead an accelerated and very ambitious program including manned spaceflight, space weapons, reconnaissance systems, large boosters, space stations, and even a lunar landing by 1967–1970. Schriever was promoted to General and took command of AFSC in July 1961.

He remained very active after retiring in 1966, and in June 1998 Falcon AFB was renamed in his honor.

Wernher von Braun
(March 23, 1912–June 16, 1977)

Dr. Wernher von Braun was the world's leading developer of missile technology, leading the teams that built the German V-2 IRBM during World War II and then the American teams that launched the first U.S. satellite and the Saturn V boosters for the Apollo Moon landings. In July 1955, von Braun proposed that Project Orbiter, a small satellite to be launched atop a V-2 derived Redstone booster, be selected for America's International Geophysical Year (IGY) program. Despite the fact that Project Orbiter was America's best and most proven booster, the IGY committee did not select it for a variety of reasons, including the fact that the Naval Research Laboratory booster selected was better suited to maintain a more civilian face on the IGY effort because it was not directly associated with any major military missile program; the committee chairman even suggested privately in 1960 that the desire to avoid having a Nazi-tainted booster lead the United States into the space age was a significant factor in the decision. At Senator Johnson's hearings after *Sputnik*, von Braun forcefully advocated that the United States create a strong civilian space agency, a position he was ideally situated to support after his team of missile engineers was transferred from ABMA to NASA in July 1960.

James E. Webb
(October 7, 1906–March 27, 1992)

James Webb was NASA Administrator from February 1961 to October 1968 and is remembered for providing strong and politically savvy leadership that contributed directly to the success of the Apollo Program. Webb was focused almost entirely on winning the Moon race but also parried attempts by DoD to gain control over human spaceflight efforts, arguing that any restructuring could delay the Apollo Program. On January 23, 1963, Webb signed an agreement to allow DoD experiments on Gemini missions. He also discussed with DoD the possibility of a joint space station project and on April 27, 1963, pledged that NASA would not initiate station development without the approval of the other agencies but never subsequently agreed to any joint station effort.

Caspar W. Weinberger
(August 18, 1917–March 28, 2006)

Caspar Weinberger served as Secretary of Defense from January 1981 to November 1987, managing President Reagan's defense buildup and setting the conditions to end the Cold War. SDI was a key element in superpower relations at this time and Weinberger's leadership was indispensable in setting the foundation for this program. An important secret meeting took place at the White House on February 11, 1983, involving Weinberger, the JCS, National Security Adviser Bud McFarlane, and President Reagan. Weinberger and the JCS provided unanimous support for investigating new strategic defense possibilities at the meeting and this was an essential step in the process leading to President Reagan's March 23, 1983, SDI speech. Weinberger provided overall supervision and direction for the major study efforts that endorsed SDI and set the organizational structure for SDIO, making it an office that reported directly to the secretary in order to insulate it from bureaucratic and budgetary pressures.

Thomas D. White
(August 6, 1901–December 22, 1965)

General Thomas D. White served as Chief of Staff of the Air Force from 1957 to 1961. In 1954, as Vice Chief of Staff, he supported ICBM development, directing Atlas be given the highest R&D priority in the Air Force and creating the Western Development Division to speed this effort. White presented the first comprehensive and official expression of Air Force space doctrine at the National Press Club on November 29, 1957, asserting that space control would lead to control over the surface of the Earth and emphasizing that the United States must control space. These assumptions about space as the new high ground became cornerstones for the high-ground doctrine and efforts to develop significant U.S. military space forces; they were also linked to the central but still controversial Air Force doctrinal tenet about the need for air superiority to enable success in all other military operations. White's second major doctrinal assumption addressed the aerospace concept: air and space form an indivisible operational medium. Stressing Air Force expertise in experimental flight in near space, White implied the logic of extending Air Force jurisdiction further out into this indivisible medium. White's presentation strongly asserted the need for space control due to the Air Force's perception of space as the ultimate high ground for future conflict, tied the service to the aerospace concept of an indivisible operational medium of air and space, and implied the Air Force was best prepared to address the grave national security challenges of space. This approach, however, put the Air Force in direct conflict with Eisenhower's policy on space as a sanctuary for the development, use, and protection of reconnaissance satellites

Jerome B. Wiesner
(May 30, 1915–October 21, 1994)

Dr. Jerome Wiesner served as Science Adviser under Presidents Kennedy and Johnson from 1961 to 1964. Wiesner headed the Ad Hoc Committee on Space and tasked this group to study the structure for and the direction of U.S. space efforts. Kennedy met with this group on January 10, 1961, and an unclassified version of their report was released the next day. The report recommended a revitalization of the NASC, called for primary emphasis on space-science missions, and warned against attempting to race the Soviets for manned space spectaculars. Wiesner strongly recommended one agency be in charge of all space-related issues. This prompted Secretary of Defense Robert McNamara to call for a review of military space organizations that led to Defense Directive 5160.32, "Development of Space Systems," issued on March 6, 1961.

6

Data and Documents

Documents

NSC-5520, "Draft Statement of Policy on U. S. Scientific Satellite Program," (Washington: National Security Council, May 27, 1955). Originally Classified Secret, Declassified on March 8, 1996.
Excerpt

Draft Statement of Policy on U.S. Scientific Satellite Program
General Considerations

1. The U. S. is believed to have the technical capability to establish successfully a small scientific satellite of the earth in the fairly near future. Recent studies by the Department of Defense have indicated that a small scientific satellite weighing 5 to 10 pounds can be launched into an orbit about the earth using adaptations of existing rocket components. If a decision to embark on such a program is made promptly, the U. S. will probably be able to establish and track such a satellite within the period 1957–58.

2. The report of the Technological Capabilities Panel of the President's Science Advisory Committee recommended that intelligence applications warrant an immediate Program leading to a very small satellite in orbit around the earth, and that re-examination should be made of the principles or practices of international law with regard to "Freedom of Space" from the standpoint of recent advances in weapon technology.

3. On April 16, 1955, the Soviet Government announced that a permanent high-level, interdepartmental Commission for interplanetary communications has been created in the Astronomical Council of the USSR Academy of Sciences. A group of Russia's top scientists is now believed to be working on a satellite program. In September 1954 the Soviet Academy of Sciences announced the establishment of the Tsoilkousky Gold Medal which would be awarded every three years for outstanding work in the field of interplanetary communications.

4. Some substantial benefits may be derived from establishing small scientific satellites. By careful observation and the analysis of actual orbital decay patterns, much information will be gained about air drag at extreme altitudes and about the fine details of the shape of and the gravitational field of the earth. Such satellites promise to provide direct and continuous determination of the total ion content of the ionosphere. These significant findings will find ready application in defense communication and missile research. When large instrumented satellites are established, a number of other kinds of scientific data may be acquired. The attached Technical Annex (Annex A) contains a further enumeration of scientific benefits.

5. From a military standpoint, the Joint Chiefs of Staff have stated their belief that intelligence applications strongly warrant the construction of a large surveillance satellite. While a small scientific satellite cannot carry surveillance equipment and therefore will have no direct intelligence potential, it does represent a technological step toward the achievement of the large surveillance satellite, and will be helpful to this end so long as the small scientific satellite program does not impede development of the large surveillance satellite.

6. Considerable prestige and psychological benefits will accrue to the nation which first is successful in launching a satellite. The inference of such a demonstration of advanced technology and its unmistakable relationship to intercontinental ballistic missile technology might have important repercussions on the political determination of free world countries to resist Communist threats, especially if the USSR were to be the first to establish a satellite. Furthermore, a small scientific satellite will provide a test of the principle of "Freedom of Space." The implications of this principle are being studied within the Executive Branch. However, preliminary studies indicate that there is no obstacle under international law to the launching of such a satellite.

7. It should be emphasized that a satellite would constitute no active military offensive threat to any country over which it might pass. Although a large satellite might conceivably serve

to launch a guided missile at a ground target, it will always be a poor choice for the purpose. A bomb could not be dropped from a satellite on a target below, because anything dropped from a satellite would simply continue alongside in the orbit.

8. The U. S. is actively collaborating in many scientific programs for the International Geophysical Year (IGY), July 1957 through December 1958. The U. S. National Committee of the IGY has requested U. S. Government support for the establishment of a scientific satellite during the Geophysical Year. The IGY affords an excellent opportunity to mesh a scientific satellite program with the cooperative world-wide geophysical observational program. The U. S. can simultaneously exploit its probable technological capability for launching a small scientific satellite to multiply and enhance the over-all benefits of the International Geophysical Year, to gain scientific prestige, and to benefit research and development in the fields of military weapons systems and intelligence. The U. S. should emphasize the peaceful purposes of the launching of such a satellite, although care must be taken as the project advances not to prejudice U. S. freedom of action (1) to proceed outside the IGY should difficulties arise in the IGY procedure, or (2) to continue with its military satellite programs directed toward the launching of a large surveillance-type satellite when feasible and desirable.

9. The Department of Defense believes that, if preliminary design studies and initial critical component development are initiated promptly, sufficient assurance of success in establishing a small scientific satellite during the IGY will be obtained before the end of this calendar year to warrant a response, perhaps qualified, to an IGY request. The satellite itself and much information as to its orbit would be public information. The means of launching would be classified.

10. A program for a small scientific satellite could be developed from existing missile programs already underway within the Department of Defense. Funds of the order of $20 million are estimated to be required to give reasonable assurance that a small scientific satellite can be established during 1957–58 (see Financial Appendix).

Courses of Action

11. Initiate a program in the Department of Defense to develop the capability of launching a small scientific satellite by 1958, with the understanding that this program will not prejudice continued research directed toward large instrumented satellites for additional research and intelligence purposes, or materially delay other major Defense programs.

12. Endeavor to launch a small scientific satellite under international auspices, such as the International Geophysical Year, in order to emphasize its peaceful purposes, provided such international auspices are arranged in a manner which: A. Preserves U.S. freedom of action in the field of satellites and related programs. B. Does not delay or otherwise impede the U.S. satellite program and related research and development programs. C. Protects the security of U.S. classified information regarding such matters as the means of launching a scientific satellite. D. Does not involve actions which imply a requirement for prior consent by any nation over which the satellite might pass in its orbit, and thereby does not jeopardize the concept of "Freedom of Space."

Financial Appendix

1. Funds of the order of $20 million are estimated to be required to assure a small scientific satellite during the period of the IGY. This figure allows for design and production of adequate vehicles and for scientific instrumentation and observation costs. It also includes preliminary back-up studies of an alternate system without vehicle procurement. The ultimate cost of a scientific satellite program will be conditioned by (1) size and complexity of the satellite, (2) longevity of each satellite, and (3) duration of the scientific observation program. Experience has shown that preliminary budget estimates on new major experimental and design programs may not anticipate many important developmental difficulties, and may therefore be considerably less than final costs.
2. The estimate of funds required is based on:

satellite vehicle	$10–$15 million
instrumentation for tracking	$2. 5 million
logistics for launching and tracking	$2. 5 million
TOTAL	$15–$20 million

3. These estimates do not include funding for military research and development already part of other missile programs. They include costs for observations that might properly be undertaken by Department of Defense agencies as part of the Department of Defense mission. They do not include costs of other observations that may be proposed by other agencies. They will provide a minimum satellite, for which two vehicle systems now under study offer good promise, "Orbiter" and "Viking." They also include exploratory studies for a back-

up program based upon the "Atlas" missile and "Aerobee"
research rocket development.

Source: United States Department of State. Foreign relations of the
United States, 1955–1957. United Nations and general international
matters. Volume XI (1955–1957). United States interest in the scientific
exploration of outer space, pp. 724–730

NSC-5814, "U.S. Policy on Outer Space," (Washington: National Security Council, June 20, 1958). Originally Classified Secret, Declassified on February 24, 1996. Excerpt

Introductory Note

*This statement of U.S. Policy on Outer Space is designated Preliminary
because man's understanding of the full implications of outer space is
only in its preliminary stages. As man develops a fuller understanding
of the new dimension of outer space, it is probable that the long-term
results of exploration and exploitation will basically affect international
and national political and social institutions.*

*Perhaps the starkest facts which confront the United States in the
immediate and foreseeable future are (1) the USSR has surpassed the
United States and the Free World in scientific and technological accomplishments in outer space, which have captured the imagination
and admiration of the world; (2) the USSR, if it maintains its present
superiority in the exploitation of outer space, will be able to use that
superiority as a means of undermining the prestige and leadership of
the United States; and (3) the USSR, if it should be the first to achieve
a significantly superior military capability in outer space, could create
an imbalance of power in favor of the Sino-Soviet Bloc and pose a direct
military threat to U.S. security.*

*The security of the United States requires that we meet these chal-
lenges with resourcefulness and vigor.*

General Considerations Introduction
Significance of Outer Space to U.S. Security

1. More than by any other imaginative concept, the mind of man
   is aroused by the thought of exploring the mysteries of outer
   space.

2. Through such exploration, man hopes to broaden his horizons, add to his knowledge, improve his way of living on earth. Already, man is sure that through further exploration he can obtain certain scientific and military values. It is reasonable for man to believe that there must be, beyond these areas, different and great values still to be discovered.

3. The technical ability to explore outer space has deep psychological implications over and above the stimulation provided by the opportunity to explore the unknown. With its hint of the possibility of the discovery of fundamental truths concerning man, the earth, the solar system, and the universe, space exploration has an appeal to deep insights within man which transcend his earthbound concerns. The manner in which outer space is explored and the uses to which it is put thus take on an unusual and peculiar significance.

4. The beginning stages of man's conquest of space have been focused on technology and have been characterized by national competition. The result has been a tendency to equate achievement in outer space with leadership in science, military capability, industrial technology, and with leadership in general.

5. The initial and subsequent successes by the USSR in launching large earth satellites have profoundly affected the belief of peoples, both in the United States and abroad, in the superiority of US. leadership in science and military capability. This psychological reaction of sophisticated and unsophisticated peoples everywhere affects U.S. relations with its allies, with the Communist Bloc, and with neutral and uncommitted nations.

6. In this situation of national competition and initial successes by the USSR, further demonstrations by the USSR of continuing leadership in outer space capabilities might, in the absence of comparable U.S. achievements in this field, dangerously impair the confidence of these peoples in U.S. over-all leadership. To be strong and bold in space technology will enhance the prestige of the United States among the peoples of the world and create added confidence in U.S. scientific, technological, industrial and military strength.

7. The novel nature of space exploitation offers opportunities for international cooperation in its peaceful aspects. It is likely that certain nations may be willing to enter into cooperative arrangements with the United States. The willingness of the Soviets to cooperate remains to be determined. The fact that the results of cooperation in certain fields, even though entered into for peaceful purposes, could have military application, may condition the extent of such cooperation is those fields.

Problem of Defining Space

10. The term *"outer space"* has no generally accepted precise
    definition.

    a. If it becomes desirable to define the lower limit of outer
       space in terms of distance from the earth's surface, three
       considerations, among others are pertinent:

       (1) *Aerodynamic Lift.* The theoretical upper limit of the
           continuous flight of winged aircraft based upon
           aerodynamic force is approximately 55 miles above the
           earth's surface.
       (2) *Satellite Orbits.* Aerodynamic drag affects the orbiting
           characteristics of earth's satellites appreciably when
           they descend to altitudes of about 100 miles and causes
           actual burning up at altitudes from 90 down to 70 miles.
           Accordingly, something over 70 miles may be considered
           the lower limit at which a satellite can exist in orbit, even
           for a short time.
       (3) *Air Defense Capability.* Today the United States has air
           defense systems which can operate against aircraft up to
           about 10 to 12 miles, which will be extended to 15 to 20
           miles in the relatively near future. Anti-missile systems
           now under development could potentially increase this
           altitude to approximately 90 to 95 miles in the foreseeable
           future. Accordingly, the upper limit of the air space which
           a nation is capable of defending might be arbitrarily set
           as lying between 90 to 95 miles above the earth's surface.

    b. Because it may not be desirable to define outer space in terms
       of distance from the earth's surface, outer space could be
       defined on a functional basis, such as in terms of the traversing
       or operating of man-made objects (at present principally
       satellites and missiles) in what is generally regarded as
       "space."

11. Although the successful orbiting of earth satellites has raised
    the question of national sovereignty *ad coelum* and as to the
    doctrine of "freedom of space," the United States has not
    recognized any upper limit to sovereignty. In order to maintain
    (a) flexibility in international negotiations in respect to all uses
    of "space," and pending a safeguarded international control
    agreement, (b) freedom of action with respect to the military
    uses of "space," the United States has taken no primary position
    on the definition of outer space.
12. Because the question of rights in "outer space" will
    undoubtedly arise at the UN General Assembly in September

1958, perhaps in international discussions on post-IGY activities, and perhaps in other international negotiations, it would appear desirable for the United States to develop a common understanding of the term "outer space" as related to particular objects and activities therein.

13. For the purposes of this policy statement, space is divided into to two regions: "air space" and "outer space." "Outer space" is considered as contiguous to "air space," with the lower limit of "outer space" being the upper limit of "air space."

Use of Outer Space
General

14. Outer space can be used:

   a. By vehicles or other objects that achieve their primary purpose in outer space; such as

   (1) Vehicles or objects that remain in an area directly over a nation's own territory, such as sounding rockets
   (2) Vehicles or objects that orbit the earth;
   (3) Vehicles that traverse outer space enroute to the moon, other planets or the sun;

   b. For the transmission of electromagnetic energy for such purposes as communications, radar measurement and electronic countermeasures;

   c. By weapons, such as ballistic missiles, and other vehicles which traverse outer space, but which achieve their primary purpose upon their return to air space or earth.

Science and Technology

17. Outer space activity and scientific research would have both military and nonmilitary applications. Examples are satellites as navigational aids; and satellites as relay stations to receive and relay television or radio signals and improve world-wide communications.

18. It is not possible to foresee all applications of outer space activity which may be developed, but our ability to achieve and maintain leadership in such applications will largely depend on the breadth of the scientific research which is undertaken and supported.

Military

19. The effective use of outer space by the United States and the Free World will enhance their military capability. Military uses of outer space (some of which may have peaceful applications) may be divided into the following three general categories:

a. *Now Planned or in Immediate Prospect*

(1) *Ballistic Missiles.* A family of IRBM's and ICBM's is now in the latter stages of development. Components of these missiles can be used to develop other space vehicles, for both military and scientific use.

(2) *Anti-ICBM's* which are now being developed.

(3) *Military Reconnaissance.* (See "Reconnaissance Satellites" section, paragraphs 20–23)

b. *Feasible in the Near Future*

(1) *Satellites of Weather Observation.*

(2) *Military Communications Satellites.*

(3) *Satellites for Electronic Countermeasures (Jamming).*

(4) *Satellites as Aids for Navigation,* tracked from the earth's surface visually or by radio.

c. *Future Possibilities*

(1) *Manned Maintenance and Resupply Outer Space Vehicles.*

(2) *Manned Defensive Outer Space Vehicles,* which might capture, destroy or neutralize an enemy outer space vehicle.

(3) *Bombardment Satellites* (Manned or Unmanned). It is conceivable that, in the future, satellites carrying weapons ready for firing on signal might be used for attacking targets on the earth.

(4) *Manned Lunar Stations,* such as military communications relay sites or reconnaissance stations. Conceivably, launching of missiles to the earth from lunar sites would be possible.

Reconnaissance Satellites

20. Reconnaissance satellites are of critical importance to U.S. national security. Those now planned are designed: (a) to gather military intelligence data, weather data, and information on the economic potential of the Sino-Soviet Bloc; and (b) to detect the launching of a missile or air attack upon the United States or its allies. Reconnaissance satellites would also have a high potential use as a means of implementing the "open skies" proposal or policing a system of international armaments control.

21. As envisaged in U.S. plans, the instrumentation of reconnaissance satellites would consist primarily of three types:

a. *Optical.* Such instrumentation will provide for earth photography and/or TV scanning. Either system might transmit by recoverable transmit techniques, or electronically, either immediately or after temporary storage. Optical instrumentation currently available for use in satellites could give resolution down to 100 feet, permitting identification of naval movements, concentrations of shipping, and possibly the general purpose of industrial complexes. Further improvement of the implementation would, through resolution much below 100 feet, make possible surveillance of airfields and concentrations of military equipment, and identification of the specific purposes of industrial complexes.

b. *Infra-red.* A satellite equipped with an infra-red sensor and operating at low satellite altitudes, could detect and track jet bombers operating above 30,000 feet and a ballistic missile from about one minute after launch to burnout. With sufficient satellites orbiting, equipped with suitable communication links, it is possible that aircraft or missile attacks launched against the United States could be detected so as to prove improved warning time to U.S. active and passive defenses (for long range missiles, up to 30 minutes).

c. *Electromagnetic.* Electromagnetic instrumentation could pick up a wide variety of electronic emanations of interest to intelligence, including location and identification of Soviet radars, frequencies, codes and jamming patterns. Data collection by satellites further offers possibilities of acquiring some line-of-sight radio and TV communications signals.

22. At some later time, when manned outer space flight is a reality, it is probable that man, in a reconnaissance satellite, could add significantly to the effectiveness of reconnaissance operations. The degree of this added effectiveness can be predicted only after more is known of unmanned reconnaissance satellite operations and of man's ability to function in an outer space environment.

23. Some political implications of the use of reconnaissance satellites may be adverse. Therefore, studies must be urgently undertaken in order to determine the most favorable political framework in which such satellites would operate.

Manned Exploration of Outer Space

24. In addition to satisfying man's urge to explore new regions, manned exploration of outer space is of importance to our national security because:

a. Although present studies in outer space can be carried on satisfactorily by using only unmanned vehicles, the

time will undoubtedly come when man's judgment
and resourcefulness will be required fully to exploit the
potentialities of outer space.

b. To the layman, manned exploration will represent the true
conquest of outer space. No unmanned experiment can
substitute for manned exploration in its psychological effect
on the peoples of the world.

c. Discovery and exploration may be required to establish a
foundation for the rejection of USSR claims to exclusive
sovereignty of other planets which may be visited by
nationals of the USSR.

25. The first step in manned outer space travel could be
undertaken using rockets and components now under study
and development. Travel by man to the moon and beyond will
probably require the development of new basic vehicles and
equipment.

Other Implications of Outer Space Activities
International Cooperation and Control
General

26. International cooperation in certain outer space activities
appears highly desirable from a scientific, political and
psychological standpoint and may appear desirable in selected
instances with U.S. allies from the military standpoint.
International cooperation agreements in which the United
States participates could have the effect of (a) enhancing the
position of the United States as a leader in advocating the
uses of outer space for peaceful purposes and international
cooperation in science, (b) conserving U.S. resources, (c)
speeding up outer space achievements by the pooling of talents,
(d) "opening up" the Soviet Bloc, and (e) introducing a degree
of order and authority in the necessary international regulations
governing certain outer space activities.

27. Various types of international cooperation may be possible
through existing international scientific organizations, the
United Nations, multilateral and bilateral arrangements with
the Free World nations and NATO, and U.S.-Soviet bilateral
arrangements. International cooperation by the United States
in outer space activities might, as consistent with U.S. security
interests, include (a) the collection and exchange of information
on outer space; (b) the exchange of scientific instrumentation;
(c) contacts among scientists; (d) participation of foreign
scientists in U.S. space projects; (e) planning and coordination
of certain programs or specific projects to be carried out on

a fully international basis (some of which might be: a large instrumented scientific satellite, communication satellites, and meteorological satellites); (f) establishment of regulations governing certain outer space activities; (g) provision and launching of scientific satellites in support of international planning of a program of satellite observations.

28. Under present conditions, the extent of international cooperation, particularly in fields having important military applications such as propulsion and guidance mechanisms, will have to take into account security considerations (see paragraphs 7 and 15).

U.S. Position

29. In January 1957 the United States initiated international discussion of the control of outer space by proposing in the UN General Assembly that the testing of outer space vehicles should be carried out and inspected under international auspices. This proposal was based on a policy decision to seek to assure that the sending of objects into outer space should be exclusively for peaceful and scientific purposes and that, under effective control, the production of such objects designed for military purposes should be prohibited as part of an armaments control system. It was thought, at the then state of the art, that a control of testing would have precluded development until more comprehensive controls could be agreed upon. The U.S. proposal was altered with the passage of time and, as presented on August 29, 1957 as the Four Power Proposal in London, calls for technical studies of the "design of an inspection system which would make it possible to assure that the sending of objects through outer space will be exclusively for peaceful and scientific purposes." In his letter of January 13, 1958 to Bulganin, the President proposed, as part of a five-point program relating to control of armaments and armed forces, that "we agree that outer space be used only for peaceful purposes" and inquired "can we not stop the productions of such weapons which would use or more accurately misuse, outer space . . . ?" In his later letter to Bulganin, dated February 15, 1958, the President proposed "wholly eliminating the newest types of weapons which use outer space for human destruction."

30.

a. The most recent statement of basic policy relating to the regulation and reduction of armed forces and armaments appears in paragraph 40 of NSC 5810/1 (May 5, 1958).

b. Further consideration of U.S. policy concerning the scope of control and inspection required to assure that outer space could be used only for peaceful purposes, as well as the relationship of any such control arrangement to other aspects of an arms agreement, is deferred pending the recommendation of the Special NSC Committee established to make preparations for a possible Summit Meeting (NSC Action No. 1893). It is understood that the Special NSC Committee will also consider possible interim and more limited arrangements, and take into account the technical feasibility of assuring that outer space can be used only for peaceful purposes.

Legal Problems of Air Space and Outer Space

33. *Numerous legal problems* will be posed by the development of activities in space. Many of these cannot be settled until we gain more experience and basic information, because the only foundation for a sound rule of law is a body of ascertained fact. It is altogether likely that some issues in the field of space law which will be practical questions in the future are not even identified today. This is not to say that there is an entire lack of international law applicable to activities in space at the present time. For example, Article 51 of the Charter of the United Nations recognizes the inherent right of individual or collective self-defense against armed attack. Clearly this right is available against any space activities employed in such an attack.

34. *International Geophysical Year.* From the arrangements and announcements made in connection with the International Geophysical Year, there may be a general implied consent that scientific satellites be launched and orbited during the IGY. Such implied consent does not necessarily mean, however, that assent has been given to the launching and orbiting or other types of satellites and missiles, or that assent with respect to scientific satellites extends beyond the IGY. It remains to be determined what rules will apply to subsequent satellites; what limitations will govern the types and purposes of satellites in the future. The United States, as well as other countries, has not yet taken positions on these questions and, here again, the answer will depend not only upon what others are likely to do but also upon what activities the United States wishes to be free to engage in.

36. *The problem of legal definitions* is unsolved. As indicated above, there is as yet insufficient basis for legally deciding that air space extends so far and no farther; that outer space begins at a given point above the earth. Because, for some time to come, at least, activities in outer space will be closely connected with

activities on the earth and in air space, many legal problems with respect to space activities may well be resolved without the necessity of determining or agreeing upon a line of demarcation between air space and outer space. If, by analogy to the Antarctic proposal of the United States, international agreement can be reached upon permissible activities in space and the rules and regulations to be followed with respect thereto, problems of sovereignty may be avoided or least deferred.

38. *Problems of national and international regulation* over activities in space will also arise. There is already the need to assign telecommunication wavelengths to communications with satellites and space objects. Other types of regulations having serious security implications will have to be worked out for the identification of space objects and for some type of traffic control to prevent congestion and interference.

39. *Generally speaking, rules will have to be evolved gradually* and pragmatically from experience. While the nations engaging in space activities will play an important role in this field, it will have to be recognized from the nature of the subject that all nations have a legitimate interest in it. The field is not suitable for abstract *a priori* codification.

Comparison of USSR and U.S. Capabilities in Outer Space Activities

40. Conclusive evidence shows that the Soviets are conducting a well-planned outer space program at high priority. The table below attempts to estimate the U.S. and USSR timetable for accomplishment of specific outer space flight activities.

   a. Soviet space flight capabilities estimated in the table reflect the earliest possible time periods in which each specific event could be successfully accomplished.

   (1) The space flight program is in competition with many other programs, particularly the missile program. The USSR probably cannot successfully accomplish all of the estimated space flight activities within the time periods specified. The USSR will not permit its space flight program to interfere with achieving an early operational capability for ICBM's (which enjoy the highest priority).

   (2) The USSR is believed to have the intention to pursue both an active space flight program designed to put man into outer space for military and/or scientific purposes, and further scientific research utilizing earth satellites, lunar rockets, and probes of Mars and Venus; but it cannot be determined, at this time, whether the basic

scientific program or the "man in space" program enjoys the higher priority and will, therefore, be pursued first.

b. U.S. space flight capabilities indicated in the table reflect the earliest possible time periods in which each specific event could be successfully accomplished. Not all of the indicated activities could be successfully accomplished with the time period specified. It must also be recognized that the accomplishment of some of the activities listed would impinge upon space activities already programmed, or upon other military programs.

41. If the USSR high-priority outer space program continues, the USSR will maintain its lead at least for the next few years, as shown in the following table.

Level of Effort

42.

b. The level of material and scientific effort to be expended on outer space activities must nevertheless be related to other national security programs to ensure that a proper balance is maintained between anticipated scientific, military and psychological gains from outer space programs and the possible loss resulting from reductions in resources allocated to other programs.

Objectives

43. The fullest development and exploitation of U.S. outer space capabilities as needed to achieve U.S. scientific, military and political purposes as follows:

a. A technological capability to meet the requirements of b, c and d below.

b. A degree of competence and a level of achievement in outer space basic and applied research and exploration which is at least on a par with that of any other nation.

c. Applications of outer space technology, research and exploration to achieve a military capability in outer space sufficient to assure the over-all superiority of U.S. [outer space] offensive and defensive systems relative to those of the USSR.

d. Applications of outer space technology, research, and exploration for nonmilitary purposes, which are at least on a par with any other nation.

    e. World recognition of the United States as, at least, the equal of any other nation in over-all outer space activity and as the leading advocate of the peaceful exploitation of outer space.

43. The establishment of the United States as the recognized leader in the over-all development and exploitation of outer space for scientific, military and political purposes.
44. As consistent with U.S. security, achievement of international cooperation in the uses of and activities related to outer space: for peaceful purposes, and with selected allies for military purposes.
45. As consistent with U.S. security, the achievement of suitable international agreements relating to the uses of outer space for peaceful purposes that will assure orderly outer space programs.
46. Utilization of the potentials of outer space to assist in "opening up" the Soviet Bloc through improved intelligence and programs of scientific cooperation.

Policy Guidance
Priority and Scope of Outer Space Effort

47. With a priority and scope sufficient to enable the United States at the earliest practicable time to achieve its scientific, military and political objectives as stated in paragraph 43, develop and expand selected U.S. activities related to outer space in:

    a. Research and technology required to exploit the military and non-military potentials of outer space.

    b. Outer space exploration required to determine such military and non-military potentials.

    c. Applications of outer space research, technology, and exploration to develop outer space capabilities (in addition to those capabilities which now have the highest national priority) required to achieve such objectives.

48. In addition to undertaking necessary immediate and short-range activities related to outer space, develop plans for outer space activities or the longer range (through at least a ten-year period).
49. Study on a continuing basis the implications which U.S. and foreign exploitation of outer space may hold for international and national political and social institutions. Critically examine such exploitation for possible consequences on activities and on life on earth (e.g., outer space activities which affect weather, health for other factors relating to activities and life on earth).
50. In the absence of a safeguarded international agreement for the control of armaments and armed forces, place primary

emphasis on activities related to outer space necessary to
maintain the over-all deterrent capability of the United States
and the Free World.

Psychological Exploitation

51. In the near future, while the USSR has a superior capability
    in space technology, judiciously select (without prejudicing
    activities under paragraph 47) projects for implementation
    which, while having scientific or military value, are designed to
    achieve a favorable world-wide psychological impact.
52. Identify, to the greatest extent possible, the interests and
    aspirations of other Free World nations in outer space with
    U.S.-sponsored activities and accomplishments.
53. Develop information and other programs that will exploit fully
    U.S. outer space activities on a continuing basis; especially,
    during the period while the USSR has superior over-all outer
    space capabilities, those designed to counter the psychological
    impact of Soviet outer space activities and to present U.S. outer
    space progress in the most favorable comparative light.

Reconnaissance Satellites

54. In anticipation of the availability of reconnaissance satellites,
    seek urgently a political framework which will place the uses
    of U.S. reconnaissance satellites in a political and psychological
    context most favorable to the United States.
55. At the earliest technologically practicable date, use recon-
    naissance satellites to enhance to the maximum extent the U.S.
    intelligence effort.

International Cooperation in Outer Space Activities

56. Consistent with the objectives in paragraphs 43 and 44, and as a
    means of maintaining the U.S. position as the leading advocate
    of the use of outer space for peaceful purposes, be prepared to
    propose that the United States join with other nations, including
    the USSR, in cooperative efforts relating to outer space.
    Specifically:

    a. Encourage a continuation and expansion of the type of
       cooperation which exists in the IGY programs, through
       non-governmental international scientific organizations
       such as the International Council of Scientific Unions;
       including cooperation in the design of experiments and
       instrumentation, exchange of information on instrumentation,
       scientific data and telemetry, exchange of instruments, and

in the use of scientific satellites and other scientific vehicles in support of international planning for exploration of outer space.

c. Invite scientists of foreign countries, including the Soviet Bloc in general on a reciprocal basis, to participate in selected U.S. programs for the scientific exploration of space.

d. Propose scientific bilateral arrangements with other nations (including the USSR) for cooperative ventures related to outer space, provided that the combined existing competence might achieve meaningful scientific and technical advance.

e. Propose to groups of nations and international organizations independent outer space projects which would be appropriate for multilateral participation.

f. Assist selected Free World nations willing and able to undertake useful activities related to outer space, [as necessary to assure that the over-all Free World position in outer space developments is at least on a par with that of the Sino-Soviet Bloc]

Limited International Arrangements to Regulate Outer Space Activities

57. Propose international agreements concerning appropriate means for maintaining a full and current public record of satellite orbits and emission frequencies.

International Outer Space Law

58. Consider as possible, the right of passage through outer space of any orbiting object that is so designed that it cannot physically interfere with the legitimate activities of other nations

59. Reserve the U.S. position on legal issues of outer space, but undertake on an urgent basis a study of the legal issues that will arise from national and international outer space activities in the near future.

Interim Position in International Negotiations

60. In negotiations with other nations or organizations dealing with outer space (pending the results of the study referred to in paragraph 58), seek to achieve common agreement to relate such negotiations to the traversing or operating of man-made objects in outer space, rather than to defined regions in outer space.

Security Classification

61. In considering whether U.S. outer space information and material requires classification under Executive Order

No. 10501, take special account of the lead achieved by the
USSR in outer space activities and the advantages, including
more rapid progress, which could accrue to the United States
through liberalizing the general availability and use of such
information and material.

Administration of Outer Space Programs

62. Provide through appropriate legislation for the conduct of U.S.
outer space activities under the direction of a civilian agency,
except in so far as such activities may be peculiar to or primarily
associated with weapons systems or military operations, in
the case of which activities the Department of Defense shall be
responsible.

Source: United States Department of State. Foreign relations of the
United States, 1958–1960. United Nations and general international
matters. Volume II (1958–1960). Outer space, pp. 845–863

National Aeronautics and Space Act of 1958 (Washington: National Archives and Records Administration, Public Law 85–568, Signed by the President on July 29, 1958). Excerpt

Title I-Short Title, Declaration of Policy, and Definitions Short Title

*Sec. 101. This act may be cited as the "National Aeronautics and Space
Act of 1958."*

NASA Act 1958
Declaration of Policy and Purpose

Sec. 102. (a) The Congress hereby declares that it is the policy of the United
States that activities in space should be devoted to peaceful purposes
for the benefit of all mankind. (b) The Congress declares that the general
welfare and security of the United States require that adequate provision be made for aeronautical and space activities. The Congress further
declares that such activities shall be the responsibility of, and shall be
directed by, a civilian agency exercising control over aeronautical and
space activities sponsored by the United States, except that activities peculiar to or primarily associated with the development of weapons systems, military operations, or the defense of the United States (including
the research and development necessary to make effective provision for
the defense of the United States) shall be the responsibility of, and shall

be directed by, the Department of Defense; and that determination as to which such agency has responsibility for and direction of any such activity shall be made by the President in conformity with section 201 (e).

(c) The aeronautical and space activities of the United States shall be conducted so as to contribute materially to one or more of the following objectives: (1) The expansion of human knowledge of phenomena in the atmosphere and space; (2) The improvement of the usefulness, performance, speed, safety, and efficiency of aeronautical and space vehicles; (3) The development and operation of vehicles capable of carrying instruments, equipment, supplies and living organisms through space; (4) The establishment of long-range studies of the potential benefits to be gained from, the opportunities for, and the problems involved in the utilization of aeronautical and space activities for peaceful and scientific purposes. (5) The preservation of the role of the United States as a leader in aeronautical and space science and technology and in the application thereof to the conduct of peaceful activities within and outside the atmosphere. (6) The making available to agencies directly concerned with national defenses of discoveries that have military value or significance, and the furnishing by such agencies, to the civilian agency established to direct and control nonmilitary aeronautical and space activities, of information as to discoveries which have value or significance to that agency; (7) Cooperation by the United States with other nations and groups of nations in work done pursuant to this Act and in the peaceful application of the results, thereof; and (8) The most effective utilization of the scientific and engineering resources of the United States, with close cooperation among all interested agencies of the United States in order to avoid unnecessary duplication of effort, facilities, and equipment.

Source: Record Group 255, National Archives and Records Administration, Washington, D.C.; available in NASA Historical Reference Collection, History Office, NASA Headquarters, Washington, D.C.

Department of Defense Directive 5160.32, "Development of Space Systems," (Washington: Department of Defense, March 6, 1961). Excerpt

March 6, 1961
Number 5160.32
Department of Defense Directive
SUBJECT Development of Space Systems

I. Purpose

This establishes policies and assigns responsibilities for research, development, test, and engineering of satellites, anti-satellites, space probes and supporting systems therefor, for all components of the Department of Defense.

II. Policy and assignment of responsibilities

A. Each military department and Department of Defense agency is authorized to conduct preliminary research to develop new ways of using space technology to perform its assigned function. The scope of such research shall be defined by the Director of Defense Research and Engineering in terms of expenditure limitations and other appropriate conditions.

B. Proposals for research and development of space programs and projects beyond the defined preliminary research stage shall be submitted to the Director of Defense Research and Engineering for review and determination as to whether such proposals, when transmitted to the Secretary of Defense, will be recommended for approval. Any such proposal will become a Department of Defense space development program or project only upon specific approval by the Secretary of Defense or the Deputy Secretary of Defense.

C. Research, development, test, and engineering of Department of Defense space development programs or projects, which are approved hereafter, will be the responsibility of the Department of the Air Force.

D. Exceptions to paragraph C, will be made by the Secretary of Defense or the Deputy Secretary of Defense only in unusual circumstances.

E. The Director of Defense Research and Engineering will maintain a current summary of approved Department of Defense space development programs and projects.

Source: Exploring the Unknown: Selected Documents in the History of the U.S. Civilian Space Program, Vol. II: External Relationships, John M. Logsdon, ed. History Office, NASA Headquarters, Washington, D.C., 314-15.

National Security Decision Memorandum 345, "Development of Anti-Satellite Capability," (Washington: National Security Council, January 18, 1977). Originally Classified Top Secret, Declassified on June 22, 2004.

The President is concerned about the increasing use by the USSR of space-based assets for direct support of their military forces. This trend, which can be expected to continue and which is typified by the Soviet use of ocean surveillance satellites to provide real-time targeting data for long-range anti-ship missiles, is substantially increasing the effectiveness of

Soviet forces. It represents a direct military threat to the combat forces of the United States. In light of these developments, the President has reassessed U.S. policy regarding acquisition of an anti-satellite capability and has decided that the Soviets should not be allowed an exclusive sanctuary in space for critical military supporting satellites.

Policy with Respect to U.S. Anti-Satellite Capabilities

The President wishes to emphasize that the United States will continue to stress international treaty obligations in space, including free use of outer space and non-interference with national technical means. However, to counter the direct military threat obligations, as well as to protect against higher level conflict situations in which the Soviets might abrogate current agreements, the President has decided that the United States should acquire a non-nuclear anti-satellite capability which could selectively nullify certain military important Soviet space systems, should that become necessary. In order to be able to use such an anti-satellite capability in a reversible, less provocative way at lower crisis thresholds, as well as to accomplish more permanent kill in high level crises and conflicts, means for both electronic nullification and physical destruction should be pursued.

U. S. Anti-Satellite Capability

The President directs that the Secretary of Defense take immediate steps toward the acquisition of non-nuclear anti-satellite capability, including means for electronic nullification as well as physical destruction.

An anti-satellite interceptor should be acquired on an expedited basis. It should be capable of destroying low altitude satellites and of nullifying a small number (6–10) of important Soviet military satellites within a period of one week.

A separate capability to electronically nullify critical Soviet military satellite at all altitudes up to synchronous should also be acquired on an urgent basis.

In order to avoid stimulating Soviet actions to counter electronically U.S. high altitude COMINT and ELINT collectors, the "fact of" a U.S. electronic anti-satellite capability should be classified and special compartmented security procedures should be used to protect the confidentiality of the existence and detailed characteristics of the program. Special procedures should also be established to review and authorize tests of electronic techniques. The "fact of" a U.S. low altitude anti-satellite interceptor should be treated as unclassified and normal security procedures applied to the program details.

Arms Control Initiatives

The President further directs the Director of the Arms Control and Disarmament Agency to identify and assess arms control initiatives that would complement development of a limited anti-satellite capability in an overall policy toward military space activities by:

1. Restricting development of high altitude anti-satellite interceptor capabilities.
2. Raising the crisis threshold for use of an anti-satellite.
3. Clarifying acts which constitute interference with space systems.

This effort should be coordinated with the Secretary of Defense, the Secretary of State, and the Director of Central Intelligence. It should not delay the acquisition actions called for in this memorandum.

Source: Box 1, National Security Decision Memoranda and Study Memoranda, Gerald R. Ford Library

Department of Defense Directive 3100.10, "Space Policy," (Washington: Department of Defense, July 9, 1999). Excerpts

4. Policy
It is DoD policy that:

4.1. Space is a medium like the land, sea, and air within which military activities shall be conducted to achieve U.S. national security objectives. The ability to access and utilize space is a vital national interest because many of the activities conducted in the medium are critical to U.S. national security and economic well-being.

4.2. Ensuring the freedom of space and protecting U.S. national security interests in the medium are priorities for space and space-related activities. U.S. space systems are national property afforded the right of passage through and operations in space without interference, in accordance with reference (a).

4.2.1. Purposeful interference with U.S. space systems will be viewed as an infringement on our sovereign rights. The U.S. may take all appropriate self-defense measures, including, if directed by the National Command Authorities (NCA), the use of force, to respond to such an infringement on U.S. rights.

4.3. The primary DoD goal for space and space-related activities is to provide operational space force capabilities to ensure that the United States has the space power to achieve its national security objectives . . . Contributing goals include sustaining a robust U.S. space industry and a strong, forward-looking technology base.

4.3.1. Space activities shall contribute to the achievement of U.S. national security objectives, in accordance with reference (a), by:

4.3.1.1. Providing support for the United States' inherent right of self-defense and defense commitments to allies and friends.

4.3.1.2. Assuring mission capability and access to space.

4.3.1.3. Deterring, warning, and, if necessary, defending against enemy attack.

4.3.1.4. Ensuring that hostile forces cannot prevent the United States' use of space.

4.3.1.5. Ensuring the United States' ability to conduct military and intelligence space and space-related activities.

4.3.1.6. Enhancing the operational effectiveness of U.S. and allied forces.

4.3.1.7. Countering, if necessary, space systems and services used for hostile purposes.

4.3.1.8. Satisfying military and intelligence requirements during peace and crisis as well as through all levels of conflict.

4.3.1.9. Supporting the activities of national policymakers, the Intelligence Community, the NCA, Combatant Commanders and the Military Services, other Federal officials, and continuity of Government operations.

4.4. Mission Areas. Capabilities necessary to conduct the space support, force enhancement, space control, and force application mission areas shall be assured and integrated into an operational space force structure that is sufficiently robust, ready, secure, survivable, resilient, and interoperable to meet the needs of the NCA, Combatant Commanders, Military Services, and intelligence users across the conflict spectrum.

4.5. Assured Mission Support. The availability of critical space capabilities necessary for executing national security missions shall be assured . . . Such support shall be considered and implemented at all stages of requirements generation, system

planning, development, acquisition, operation, and support.
Assured mission capability shall be assessed and taken into
account in determining tradeoffs among cost, performance,
resilience, lifetime, protection, survivability, and related factors.
Access to space, robust satellite control, effective surveillance of
space, timely constellation replenishment/reconstitution, space
system protection, and related information assurance, access
to critical electromagnetic frequencies, critical asset protection,
critical infrastructure protection, force protection, and
continuity of operations shall be ensured to satisfy the needs of
the NCA, Combatant Commanders, Military Services, and the
intelligence users across the conflict spectrum.

4.6. Planning. Planning for space and space-related activities shall
focus on improving the conduct of national security space
operations, assuring mission support, and enhancing support to
military operations and other national security objectives. Such
planning shall also identify missions, functions, and tasks that
could be performed more efficiently and effectively by space
forces than terrestrial alternatives.

4.6.1. Long-range planning objectives for space capabilities
are to:

4.6.1.1. Ensure U.S. leadership through revolutionary
technological approaches in critical areas.

4.6.1.2. Develop a responsive, customer-focused
architecture that simplifies operations and use.

4.6.1.3. Ensure civil and commercial capabilities are
used to the maximum extent feasible and
practical (including the use of allied and friendly
capabilities, as appropriate), consistent with
national security requirements.

4.6.1.4. Provide assured, cost-effective, responsive access
to space.

4.6.1.5. Contribute to a comprehensive command,
control, communications, intelligence,
surveillance, and reconnaissance architecture
that integrates space, airborne, land, and
maritime assets.

4.6.1.6. Ensure space systems are seamlessly integrated
within a globally accessible and secure
information infrastructure.

4.6.1.7. Provide appropriate national security space
services and information to the intelligence,
civil, commercial, scientific, and international
communities.

4.6.1.8. Provide space control capabilities consistent with Presidential policy as well as U.S. and applicable international law.

4.6.1.9. Protect national security space systems to ensure mission execution.

4.6.1.10. Explore force application concepts, doctrine, and technologies consistent with Presidential policy as well as U.S. and applicable international law.

4.6.1.11. Promote a trained, space-literate national security workforce able to utilize fully space capabilities for the full spectrum of national security operations.

4.6.2. Architectures. An integrated national security space architecture, including space, ground, and communications link segments, as well as user interfaces and equipment, shall be developed to the maximum extent feasible. Such an integrated architecture shall address defense and intelligence missions and activities to eliminate unnecessary vertical stove-piping of programs, minimize unnecessary duplication of missions and functions, achieve efficiencies in acquisition and future operations, provide strategies for transitioning from existing architectures, and thereby improve support to military operations and other national security objectives.

4.6.2.1. Space architectures shall be structured to take full advantage, as appropriate, of defense, intelligence, civil, commercial, allied, and friendly space capabilities. Such architectures shall also include, as appropriate, system, operational, and technical architecture descriptions. Joint technical standards drawn from widely accepted commercial standards, consistent with national security requirements, shall provide the basis for new system integration where appropriate. Appropriate interoperability and standards mandates shall be observed to enable the interoperability of space services.

4.6.2.2. Space architectures should be designed for appropriate levels of mission optimization, availability, and survivability in all aspects of on-orbit configurations and associated infrastructure. Planning shall emphasize the

need for responsiveness and the elimination
of vulnerabilities that could prevent mission
accomplishment.

4.7. Augmentation. Requirements, arrangements, and procedures,
including cost sharing and reciprocity arrangements, for
augmentation of the space force structure by civil, commercial,
allied, and friendly space systems shall be identified in
coordination with the Director of Central Intelligence, as
appropriate . . .

4.9. Support to Commercial Space Activities. Stable and predictable
U.S. private sector access to appropriate DoD space-related
hardware, facilities, and data shall be facilitated consistent with
national security requirements, in accordance with references
(a) and (k). The U.S. Government's right to use such hardware,
facilities, and data on a priority basis to meet national security
and critical civil sector requirements shall be preserved.

4.10.1. Cost as an Independent Variable. Cost, as an
independent variable, shall be applied in all
architecture development processes to ensure requiring
organizations understand cost drivers and weigh all
requirements against their associated costs.

4.10.2. Acquisition. Acquisition strategies shall usually include:
an overview of the system's capabilities and concept of
operations desired for the full system; a flexible overall
architecture, which includes a process for change; an
emphasis on open systems design, flexible technology
insertion, and rigorous technology demonstrations;
rapid achievement of incremental capability in response
to time-phased statements of operational requirements;
and close and frequent communications with users. At
program initiation, the acquisition strategy submitted
for the cognizant acquisition authority's approval
shall describe whether an evolutionary approach is
appropriate, and, if so, how the program manager will
implement the approach. Progression to an additional
level of capability beyond the first increment requires
the cognizant acquisition authority's approval and shall
be based on a review of evolving requirements and
technology development.

4.10.3. Preference for Commercial Acquisition. Lengthy mission
specifications shall be balanced against opportunities
for technology insertion, taking into consideration
commercial-off-the-shelf solutions for national security
items, non-developmental items, and national security
adaptations of commercial items. Acquisition of national

security-unique systems shall not be authorized, in general, unless suitable and adaptable commercial alternatives are not available. Such cooperation should be based on the principles of reciprocity and tangible mutual benefits and should be pursued in a manner that reasonably protects and balances U.S. national security and economic interests.

4.10.4. Science and Technology. Leading-edge technologies that address identified mission area deficiencies shall be investigated. Investments for such technology shall feature a suitable mix of theoretical research and scientific exploration and applications which support the joint vision for military operations and other national security objectives.

4.10.5. Demonstration and Experimentation. Technology applications that address mission area deficiencies shall be demonstrated. Such demonstrations shall involve both the developmental and operational elements of the DoD Components and shall be pursued to identify the value of emerging technology to the warfighter and the national security community.

4.10.6. Research and Development. Commercial systems and technologies shall be leveraged and exploited whenever possible. Research and development investments shall focus on unique national security requirements which have no known potential, or insufficient potential, for civil or commercial sector exploitation or which require protection from disclosure. Forecasts of long-term needs shall guide investments using sound business criteria to ensure they have reasonable internal rates of return compared with alternatives.

4.10.11. Outsourcing and Privatization. Opportunities to outsource or privatize space and space-related functions and tasks, which could be performed more efficiently and effectively by the private sector, shall be investigated aggressively, consistent with the need to protect national security and public safety. Clear lines of accountability to Combatant Commanders shall be demonstrated and documented in the employment of such resources.

4.11. Operations. Space capabilities shall be operated and employed to: assure access to and use of space; deter and, if necessary, defend against hostile actions; ensure that hostile forces cannot

prevent U.S. use of space; ensure the United States' ability
to conduct military and intelligence space and space-related
activities; enhance the operational effectiveness of U.S., allied,
and friendly forces; and counter, when directed, space systems
and services used for hostile purposes.

4.11.1. Integration. Space capabilities and applications shall
be integrated into the strategy, doctrine, concepts
of operations, education, training, exercises, and
operations and contingency plans of U.S. military
forces. Space support to the lowest appropriate level,
including the lowest tactical level, shall be emphasized
and optimized to ensure that all echelons of command
understand and exploit fully the operational advantages
which space systems provide, understand their
operational limitations, and effectively use space
capabilities for joint and combined operations.

4.11.2. Education, Training, and Exercises. Information about
space force structure, missions, capabilities, and
applications shall be incorporated into Professional
Military Education as well as Joint and Service training
and exercises to provide appropriately educated
and trained personnel to all levels of joint and
component military staffs and forces. Space missions
and capabilities, the ability to operate under foreign
surveillance or against an adversary enhanced by space
capabilities, and the ability to compensate for capability
loss shall be integrated into appropriate Joint and
Service exercises.

4.11.4. Military Personnel-in-Space. The unique capabilities
that can be derived from the presence of humans in
space may be utilized to the extent feasible and practical
to perform in-space research, development, testing,
and evaluation as well as enhance existing and future
national security space missions. This may include
exploration of military roles for humans in space
focusing on unique or cost-effective contributions to
operational missions.

4.11.5. Space Debris. The creation of space debris shall be
minimized . . . Design and operation of space tests,
experiments, and systems shall strive to minimize or
reduce the accumulation of such debris consistent with
mission requirements and cost effectiveness.

4.11.6. Spacecraft End-of-Life. Spacecraft disposal at the end
of mission life shall be planned for programs involving
on-orbit operations. Spacecraft disposal shall be

accomplished by atmospheric reentry, direct retrieval, or maneuver to a storage orbit to minimize or reduce the impact on future space operations.

4.11.7. Spaceflight Safety. All DoD activities to, in, through, or from space, or aimed above the horizon with the potential to inadvertently and adversely affect satellites or humans in space, shall be conducted in a safe and responsible manner that protects space systems, their mission effectiveness, and humans in space, consistent with national security requirements. Such activities shall be coordinated with U.S. Space Command, as appropriate, for predictive avoidance or deconfliction with U.S., friendly, and other space operations.

4.12. Intersector Cooperation. Enhanced cooperation with the intelligence, civil, and commercial space sectors shall be pursued to ensure that all U.S. space sectors benefit from the space technologies, facilities, and support services available to the nation. Such cooperation shall share or reduce costs, minimize redundant capabilities, minimize duplication of missions and functions, achieve efficiencies in acquisition and future operations, improve support to military operations, and sustain a robust U.S. space industry and a strong, forward-looking space technology base. Improvement of the coordination and, as appropriate, integration of defense and intelligence space activities shall be a priority. Procedures shall be established for the timely transfer of DoD-developed space technology to the private sector consistent with the need to protect national security . . .

4.13. International Cooperation. International cooperation and partnerships in space activities shall be pursued with the United States' allies and friends to the maximum extent feasible . . . Such cooperation shall forge closer security ties with U.S. allies and friends, enhance mutual and collective defense capabilities, and strengthen U.S. economic security. It shall also strengthen alliance structures, improve interoperability between U.S. and allied forces, and enable them to operate in a combined environment in a more efficient and effective manner. Such cooperation shall be based on the principles of reciprocity and tangible, mutual benefit and shall take into consideration U.S. equities from a broad foreign policy perspective. Such cooperation shall be pursued in a manner, which protects both U.S. national security and economic security and is consistent with U.S. arms control, nonproliferation, export control, and foreign policies.

4.14. Intelligence Support. A high priority shall be placed on the collection, analysis, and timely dissemination of intelligence information to support space and space-related policy-making, requirements generation, research, development, testing, evaluation, acquisition, operations, and employment. Requirements for such intelligence support shall be identified, prioritized, and submitted through established processes to produce timely, useful intelligence products . . .

4.15. Arms Control and Related Activities. Space and space-related activities shall comply with applicable presidential policies as well as applicable domestic and international law. Space forces planning shall include the provision of appropriate responses to possible breakouts from existing arms control treaties and agreements. The President shall be advised on the military significance of potential space arms control agreements and other related measures being considered for international implementation. Positions and policies regarding arms control and related activities shall preserve the rights of the United States to conduct research, development, testing, and operations in space for military, intelligence, civil, and commercial purposes . . .

4.16. Nonproliferation and Export Controls. The Missile Technology Control Regime is the primary tool of U.S. missile nonproliferation policy . . . Space systems, technology, and information that could be used in a manner detrimental to U.S. national security interests shall be protected. Measures shall be taken to protect technologies, methodologies, information, and overall system capabilities and vulnerabilities, which sustain advantages in space capabilities and continued technological advancements. Measures shall also be taken to maintain appropriate controls over those technologies, methodologies, information, and capabilities, which could be sold or transferred to foreign recipients. Other countries' practices, U.S. foreign policy objectives, and encouragement of free and fair trade in commercial space activities shall be taken into account when considering whether to enter into space-related agreements.

E2. Enclosure 2
Definitions

E2.1.1. *Force Application.* Combat operations in, through, and from space to influence the course and outcome of conflict. The force application mission area includes: ballistic missile defense and force projection.

E2.1.2. *Force Enhancement.* Combat support operations to improve the effectiveness of military forces as well as support other intelligence, civil, and commercial users. The force enhancement mission area includes: intelligence, surveillance, and reconnaissance; tactical warning and attack assessment; command, control, and communications; position, velocity, time, and navigation; and environmental monitoring.

E2.1.3. *Space Control.* Combat and combat support operations to ensure freedom of action in space for the United States and its allies and, when directed, deny an adversary freedom of action in space. The space control mission area includes: surveillance of space; protection of U.S. and friendly space systems; prevention of an adversary's ability to use space systems and services for purposes hostile to U.S. national security interests; negation of space systems and services used for purposes hostile to U.S. national security interests; and directly supporting battle management, command, control, communications, and intelligence.

E2.1.4. *Space Forces.* The space and terrestrial systems, equipment, facilities, organizations, and personnel necessary to access, use, and, if directed, control space for national security.

E2.1.5. *Space Power.* The total strength of a nation's capabilities to conduct and influence activities to, in, through, and from the space medium to achieve its objectives.

E2.1.6. *Space Superiority.* The degree of dominance in space of one force over another, which permits the conduct of operations by the former and its related land, sea, air, and space forces at a given time and place without prohibitive interference by the opposing force.

E2.1.7. *Space Support.* Combat service support operations to deploy and sustain military and intelligence systems in space. The space support mission area includes launching and deploying space vehicles, maintaining and sustaining spacecraft on-orbit, and deorbiting and recovering space vehicles, if required.

E2.1.8. *Space Systems.* All of the devices and organizations forming the space network. These consist of: spacecraft; mission package(s); ground stations; data links among spacecraft, ground stations, mission or user terminals, which may include initial reception, processing, and exploitation; launch systems; and directly related supporting infrastructure, including space surveillance and battle management/command, control, communications, and computers.

Source: Defense Technical Information Center, Department of Defense Directive 3100.10, Space Policy, available from http://www.dtic.mil/whs/directives/corres/pdf/310010p.pdf.

"Fact Sheet: U.S. Commercial Remote Sensing Policy," (Washington: The White House, April 25, 2003). Excerpts

II. Policy Goal

The fundamental goal of this policy is to advance and protect U.S. national security and foreign policy interests by maintaining the nation's leadership in remote sensing space activities, and by sustaining and enhancing the U.S. remote sensing industry. Doing so will also foster economic growth, contribute to environmental stewardship, and enable scientific and technological excellence.

In support of this goal, the United States Government will:

- Rely to the maximum practical extent on U.S. commercial remote sensing space capabilities for filling imagery and geospatial needs for military, intelligence, foreign policy, homeland security, and civil users;
- Focus United States Government remote sensing space systems on meeting needs that can not be effectively, affordably, and reliably satisfied by commercial providers because of economic factors, civil mission needs, national security concerns, or foreign policy concerns;
- Develop a long-term, sustainable relationship between the United States Government and the U.S. commercial remote sensing space industry;
- Provide a timely and responsive regulatory environment for licensing the operations and exports of commercial remote sensing space systems; and
- Enable U.S. industry to compete successfully as a provider of remote sensing space capabilities for foreign governments and foreign commercial users, while ensuring appropriate measures are implemented to protect national security and foreign policy.

III. Background

Vital national security, foreign policy, economic, and civil interests depend on the United States ability to remotely sense Earth from space. Toward these ends, the United States Government develops and operates highly capable remote sensing space systems for national security purposes, to satisfy civil mission needs, and to provide important public services. United States national security systems are valuable assets because of their high quality data collection, timeliness, volume, and coverage that provide a near real-time capability for regularly monitoring events around the world. United States civil remote sensing systems enable such activities as research on local, regional, and global change, and support services and data products for weather, climate, and hazard response, and agricultural, transportation, and infrastructure planning.

A robust U.S. commercial remote sensing space industry can augment and potentially replace some United States Government capabilities and can contribute to U.S. military, intelligence, foreign policy, homeland security, and civil objectives, as well as U.S. economic competitiveness. Continued development and advancement of U.S. commercial remote sensing space capabilities also is essential to sustaining the nation's advantage in collecting information from space. Creating a robust U.S. commercial remote sensing industry requires enhancing the international competitiveness of the industry.

IV. Licensing and Operation Guidelines for Private Remote Sensing Space Systems

The Secretary of Commerce, through the National Oceanic and Atmospheric Administration (NOAA), licenses and regulates the U.S. commercial remote sensing space industry, pursuant to the Land Remote Sensing Policy Act of 1992, as amended, and other applicable legal authorities. The Secretary of Defense and the Secretary of State are responsible for determining the conditions necessary to protect national security and foreign policy concerns, respectively. NOAA, in coordination with other affected agencies and in consultation, as appropriate, with industry, will develop, publish, and periodically review the licensing regulations and associated timelines governing U.S. commercial remote sensing space systems.

To support the goals of this policy, U.S. companies are encouraged to build and operate commercial remote sensing space systems whose operational capabilities, products, and services are superior to any current or planned foreign commercial systems. However, because of the potential value of its products to an adversary, the operation of a U.S. commercial remote sensing space system requires appropriate security measures to address U.S. national security and foreign policy concerns. In such cases, the United States Government may restrict operations of the commercial systems in order to limit collection and/or dissemination of certain data and products, e.g., best resolution, most timely delivery, to the United States Government, or United States Government approved recipients.

On a case-by-case basis, the United States Government may require additional controls and safeguards for U.S. commercial remote sensing space systems potentially including them as conditions for United States Government use of those capabilities. These controls and safeguards shall include, but not be limited to: (1) the unique conditions associated with United States Government use of commercial remote sensing space systems; and (2) satellite, ground station, and communications link protection measures to allow the United States Government to rely on these systems. The United States Government also may condition the operation of U.S. commercial remote sensing space systems to ensure appropriate measures are implemented to protect U.S. national security and foreign policy interests.

V. United States Government Use of Commercial Remote Sensing Space Capabilities

To support the goals of this policy, the United States Government shall utilize U.S. commercial remote sensing space capabilities to meet imagery and geospatial needs. Foreign commercial remote sensing space capabilities, including but not limited to imagery and geospatial products and services, may be integrated in United States Government imagery and geospatial architectures, consistent with national security and foreign policy objectives.

VI. Foreign Access to U.S. Commercial Remote Sensing Space Capabilities

It is in U.S. national security, foreign policy, and economic interests that U.S. industry compete successfully as providers of remote sensing space products and capabilities to foreign governments and foreign commercial users. Therefore, license applications for U.S. commercial remote sensing space exports shall be considered favorably to the extent permitted by existing law, regulations and policy when such exports support these interests.

The United States Government will consider remote sensing exports on a case-by-case basis. These exports will continue to be licensed pursuant to the United States Munitions List or the Commerce Control List, as appropriate, and in accordance with existing law and regulations. The following guidance will also apply, when considering license applications for remote sensing exports:

- The United States Government will take into account exports' potential contribution to achieving the goals of this policy, the overall relationship, particularly the existing defense and defense trade relationship with the proposed recipient nation, and broader U.S. national security, foreign policy, and economic objectives;
- As a general guideline, remote sensing exports that are currently available or are planned to be available in the global marketplace also will be considered favorably;
- Exports of sensitive or advanced information, systems, technologies, and components, however, will be approved only rarely, on a case-by-case basis. These items include systems engineering and systems integration capabilities and techniques, or enabling components or technologies, i.e., items with capabilities significantly better than those achievable by current or near-term foreign systems. The Secretary of State, in consultation with the Secretary of Defense and the Director

of Central Intelligence, shall maintain a Sensitive Technology List that includes these items. This list shall be made available to U.S. industry, consistent with national security and foreign policy concerns. The Department of State shall use the list in the evaluation of requests for exports; and

- Sensitive or advanced remote sensing exports, including but not limited to a sub-set of items specifically identified on the Sensitive Technology List, will be approved only on the basis of a government-to-government agreement or other acceptable arrangement that includes, among other things, end-use and retransfer assurances that protect U.S. controlled technical data, and broader national security and foreign policy needs. Such agreements also may include protections for intellectual property and economic interests. To facilitate timely implementation, the disposition of export license applications will be expedited after completion of such agreements or arrangements.

VII. Government-to-Government Intelligence, Defense, and Foreign Relationships

The United States Government will use U.S. commercial remote sensing space capabilities to the maximum extent practicable to foster foreign partnerships and cooperation, and foreign policy objectives, consistent with the goals of this policy and with broader national security objectives. Proposals for new partnerships regarding remote sensing that would raise questions about United States Government competition with the private sector shall be submitted for interagency review. In general, the United States Government should not pursue such partnerships if they would compete with the private sector, unless there is a compelling national security or foreign policy reason for doing so.

Source: Office of Science and Technology Policy, "U.S. Commercial Remote Sensing Policy Fact Sheet, Washington D.C., April 25, 2003; available from http://www.au.af.mil/au/awc/awcgate/space/2003remotesensing-ostp.htm.

"Fact Sheet: U.S. Space-Based Positioning, Navigation, and Timing Policy," (Washington: The White House, December 15, 2004). Excerpts

II. Background

Over the past decade, the Global Positioning System has grown into a global utility whose multi-use services are integral to U.S. national se-

curity, economic growth, transportation safety, and homeland security, and are an essential element of the worldwide economic infrastructure. In the year 2000, the United States recognized the increasing impor-tance of the Global Positioning System to civil and commercial users by discontinuing the deliberate degradation of accuracy for non-military signals, known as Selective Availability. Since that time, commercial and civil applications of the Global Positioning System have continued to multiply and their importance has increased significantly. Services dependent on Global Positioning System information are now an engine for economic growth, enhancing economic development, and improving safety of life, and the system is a key component of multiple sectors of U.S. critical infrastructure.

While the growth in civil and commercial applications continues, the positioning, navigation, and timing information provided by the Global Positioning System remains critical to U.S. national security, and its applications are integrated into virtually every facet of U.S. mili-tary operations. United States and allied military forces will continue to rely on the Global Positioning System military services for positioning, navigation, and timing services.

The continuing growth of services based on the Global Positioning System presents opportunities, risks, and threats to U.S. national, home-land, and economic security. The widespread and growing dependence on the Global Positioning System of military, civil, and commercial systems and infrastructures has made many of these systems inherently vulnerable to an unexpected interruption in positioning, navigation, and/or timing services. In addition, whether designed for military capabilities or not, all positioning, navigation, and timing signals from space and their aug-mentations provide inherent capabilities that can be used by adversaries, including enemy military forces and terrorist groups. Finally, emerging foreign space-based positioning, navigation, and timing services could en-hance or undermine the future utility of the Global Positioning System.

The United States must continue to improve and maintain the Global Positioning System, augmentations, and backup capabilities to meet growing national, homeland, and economic security requirements, for civil requirements, and to meet commercial and scientific demands. In parallel, we must continue to improve capabilities to deny adversary access to all space-based positioning, navigation, and timing services, particularly including services that are openly available and can be read-ily used by adversaries and/or terrorists to threaten the security of the United States. In addition, the diverse requirements for and multiple applications of space-based positioning, navigation, and timing services require stable yet adaptable policies and management mechanisms. The existing management mechanisms for the Global Positioning System and its augmentations must be modified to accommodate a multi-use approach to program planning, resource allocation, system development, and operations. Therefore, the United States Government must improve the policy and management framework governing the Global Positioning System and its augmentations to support their continued ability to meet increasing and varied domestic and global requirements.

III. Goals and Objectives

The fundamental goal of this policy is to ensure that the United States maintains space-based positioning, navigation, and timing services, augmentation, back-up, and service denial capabilities that: (1) provide uninterrupted availability of positioning, navigation, and timing services; (2) meet growing national, homeland, economic security, and civil requirements, and scientific and commercial demands; (3) remain the pre-eminent military space-based positioning, navigation, and timing service; (4) continue to provide civil services that exceed or are competitive with foreign civil space-based positioning, navigation, and timing services and augmentation systems; (5) remain essential components of internationally accepted positioning, navigation, and timing services; and (6) promote U.S. technological leadership in applications involving space-based positioning, navigation, and timing services. To achieve this goal, the United States Government shall:

- Provide uninterrupted access to U.S. space-based global, precise positioning, navigation, and timing services for U.S. and allied national security systems and capabilities through the Global Positioning System, without being dependent on foreign positioning, navigation, and timing services;
- Provide on a continuous, worldwide basis civil space-based, positioning, navigation, and timing services free of direct user fees for civil, commercial, and scientific uses, and for homeland security through the Global Positioning System and its augmentations, and provide open, free access to information necessary to develop and build equipment to use these services;
- Improve capabilities to deny hostile use of any space-based positioning, navigation, and timing services, without unduly disrupting civil and commercial access to civil positioning, navigation, and timing services outside an area of military operations, or for homeland security purposes;
- Improve the performance of space-based positioning, navigation, and timing services, including more robust resistance to interference for, and consistent with, U.S. and allied national security purposes, homeland security, and civil, commercial, and scientific users worldwide;

- Maintain the Global Positioning System as a component of multiple sectors of the U.S. Critical Infrastructure, consistent with Homeland Security Presidential Directive-7, Critical Infrastructure Identification, Prioritization, and Protection, dated December 17, 2003;
- Encourage foreign development of positioning, navigation, and timing services and systems based on the Global Positioning System. Seek to ensure that foreign space-based positioning, navigation, and timing systems are interoperable with the civil services of the Global Positioning System and its augmentations in order to benefit civil, commercial, and scientific users worldwide. At a minimum, seek to ensure that foreign systems are compatible with the Global Positioning System and its augmentations and address mutual security concerns with foreign providers to prevent hostile use of space-based positioning, navigation, and timing services; and
- Promote the use of U.S. space-based positioning, navigation, and timing services and capabilities for applications at the Federal, State, and local level, to the maximum practical extent.

IV. Management of Space-Based Positioning, Navigation, and Timing Services

This policy establishes a permanent National Space-Based Positioning, Navigation, and Timing Executive Committee. The Executive Committee will be co-chaired by the Deputy Secretaries of the Department of Defense and the Department of Transportation or by their designated representatives. Its members will include representatives at the equivalent level from the Departments of State, Commerce, and Homeland Security, the Joint Chiefs of Staff, the National Aeronautics and Space Administration, and from other Departments and Agencies as required. Components of the Executive Office of the President, including the Office of Management and Budget, the National Security Council staff, the Homeland Security Council staff, the Office of Science and Technology Policy, and the National Economic Council staff, shall participate as observers to the Executive Committee. The Chairman of the Federal Communications Commission shall be invited to participate on the Executive Committee as a Liaison. The Executive Committee shall meet at least twice each year. The Secretaries of Defense and Transportation shall develop the procedures by which the Committee shall operate.

V. Foreign Access to U.S. Space-Based Positioning, Navigation, and Timing Capabilities

Any exports of U.S. positioning, navigation, and timing capabilities covered by the United States Munitions List or the Commerce Control List will continue to be licensed pursuant to the International Traffic in Arms Regulations or the Export Administration Regulations, as appropriate, and in accordance with all existing laws and regulations.

As a general guideline, export of civil or other non-United States Munitions List space-based positioning, navigation and timing capabilities that are currently available or are planned to be available in the global marketplace will continue to be considered favorably. Exports of sensitive or advanced positioning, navigation, and timing information, systems, technologies, and components will be considered on a case-by-case basis in accordance with existing laws and regulations, as well as relevant national security and foreign policy goals and considerations. In support of such reviews, the Secretary of State, in consultation with the Secretaries of Defense, Commerce, and Energy, the Administrator of the National Aeronautics and Space Administration, and the Director of Central Intelligence, shall modify and maintain the Sensitive Technology List directed in U.S. Commercial Remote Sensing Space Policy, dated April 25, 2003, including those technology items or areas deemed sensitive for positioning, navigation and timing applications. The Secretaries of State and Commerce shall use the list in the evaluation of requests for exports.

Source: Office of Science and Technology Policy, "U.S. Space-Based Positioning, Navigation, and Timing Policy Fact Sheet, Washington D.C., December 15, 2004; available from http://www.whitehouse.gov/files/documents/ostp/Issues/FactSheetSPACE-BASEDPOSITIONINGNAVI GATIONTIMING.pdf.

"National Space Policy of the United States of America," (Washington: The White House, 28 June 2010). Excerpts

The space age began as a race for security and prestige between two superpowers. The opportunities were boundless, and the decades that followed have seen a radical transformation in the way we live our daily lives, in large part due to our use of space. Space systems have taken us to other celestial bodies and extended humankind's horizons back in time to the very first moments of the universe and out to the galaxies at its far reaches. Satellites contribute to increased transparency and stability among nations and provide a vital communications path for avoiding potential conflicts. Space systems increase our knowledge in many scientific fields, and life on Earth is far better as a result.

The utilization of space has created new markets; helped save lives by warning us of natural disasters, expediting search and rescue operations, and making recovery efforts faster and more effective; made agriculture and natural resource management more efficient and sustainable; expanded our frontiers; and provided global access to advanced medicine, weather forecasting, geospatial information, financial operations, broadband and other communications, and scores of other activities worldwide. Space systems allow people and governments around the world to see with clarity, communicate with certainty, navigate with accuracy, and operate with assurance.

The legacy of success in space and its transformation also presents new challenges. When the space age began, the opportunities to use space were limited to only a few nations, and there were limited consequences for irresponsible or unintentional behavior. Now, we find ourselves in a world where the benefits of space permeate almost every facet of our lives. The growth and evolution of the global economy has ushered in an ever-increasing number of nations and organizations using space. The now-ubiquitous and interconnected nature of space capabilities and the world's growing dependence on them mean that irresponsible acts in space can have damaging consequences for all of us. For example, decades of space activity have littered Earth's orbit with debris; and as the world's space-faring nations continue to increase activities in space, the chance for a collision increases correspondingly.

As the leading space-faring nation, the United States is committed to addressing these challenges. But this cannot be the responsibility of the United States alone. All nations have the right to use and explore space, but with this right also comes responsibility. The United States, therefore, calls on all nations to work together to adopt approaches for responsible activity in space to preserve this right for the benefit of future generations.

From the outset of humanity's ascent into space, this Nation declared its commitment to enhance the welfare of humankind by cooperating with others to maintain the freedom of space.

The United States hereby renews its pledge of cooperation in the belief that with strengthened international collaboration and reinvigorated U.S. leadership, all nations and peoples—space-faring and space-benefiting—will find their horizons broadened, their knowledge enhanced, and their lives greatly improved.

Principles

In this spirit of cooperation, the United States will adhere to, and proposes that other nations recognize and adhere to, the following principles:

- It is the shared interest of all nations to act responsibly in space to help prevent mishaps, misperceptions, and mistrust. The United States considers the sustainability, stability, and free access to, and use of, space vital to its national interests. Space operations should be conducted in ways that emphasize openness and transparency to improve public awareness of the activities of government, and enable others to share in the benefits provided by the use of space.

- A robust and competitive commercial space sector is vital to continued progress in space. The United States is committed to encouraging and facilitating the growth of a U.S. commercial space sector that supports U.S. needs, is globally competitive, and advances U.S. leadership in the generation of new markets and innovation-driven entrepreneurship.
- All nations have the right to explore and use space for peaceful purposes, and for the benefit of all humanity, in accordance with international law. Consistent with this principle, "peaceful purposes" allows for space to be used for national and homeland security activities.
- As established in international law, there shall be no national claims of sovereignty over outer space or any celestial bodies. The United States considers the space systems of all nations to have the rights of passage through, and conduct of operations in, space without interference. Purposeful interference with space systems, including supporting infrastructure, will be considered an infringement of a nation's rights.
- The United States will employ a variety of measures to help assure the use of space for all responsible parties, and, consistent with the inherent right of self-defense, deter others from interference and attack, defend our space systems and contribute to the defense of allied space systems, and, if deterrence fails, defeat efforts to attack them.

Goals

Consistent with these principles, the United States will pursue the following goals in its national space programs:

- **Energize competitive domestic industries** to participate in global markets and advance the development of: satellite manufacturing; satellite-based services; space launch; terrestrial applications; and increased entrepreneurship.
- **Expand international cooperation** on mutually beneficial space activities to: broaden and extend the benefits of space; further the peaceful use of space; and enhance collection and partnership in sharing of space-derived information.
- **Strengthen stability in space** through: domestic and international measures to promote safe and responsible

operations in space; improved information collection and
sharing for space object collision avoidance; protection of
critical space systems and supporting infrastructures, with
special attention to the critical interdependence of space and
information systems; and strengthening measures to mitigate
orbital debris.

- **Increase assurance and resilience of mission-essential
  functions** enabled by commercial, civil, scientific, and national
  security spacecraft and supporting infrastructure against
  disruption, degradation, and destruction, whether from
  environmental, mechanical, electronic, or hostile causes.
- **Pursue human and robotic initiatives** to develop innovative
  technologies, foster new industries, strengthen international
  partnerships, inspire our Nation and the world, increase
  humanity's understanding of the Earth, enhance scientific
  discovery, and explore our solar system and the universe
  beyond.
- **Improve space-based Earth and solar observation**
  capabilities needed to conduct science, forecast terrestrial and
  near-Earth space weather, monitor climate and global change,
  manage natural resources, and support disaster response and
  recovery.

Intersector Guidelines

In pursuit of this directive's goals, all departments and agencies shall
execute the following guidance:

Foundational Activities and Capabilities

- **Strengthen U.S. Leadership In Space-Related Science,
  Technology, and Industrial Bases.** Departments and agencies
  shall: conduct basic and applied research that increases
  capabilities and decreases costs, where this research is best
  supported by the government; encourage an innovative and
  entrepreneurial commercial space sector; and help ensure the
  availability of space-related industrial capabilities in support of
  critical government functions.
- **Enhance Capabilities for Assured Access To Space.** United
  States access to space depends in the first instance on launch
  capabilities. United States Government payloads shall be
  launched on vehicles manufactured in the United States unless
  exempted by the National Security Advisor and the Assistant
  to the President for Science and Technology and Director of
  the Office of Science and Technology Policy, consistent with
  established interagency standards and coordination guidelines.

Where applicable to their responsibilities departments and agencies shall:

Work jointly to acquire space launch services and hosted payload arrangements that are reliable, responsive to United States Government needs, and cost-effective;
Enhance operational efficiency, increase capacity, and reduce launch costs by investing in the modernization of space launch infrastructure; and
Develop launch systems and technologies necessary to assure and sustain future reliable and efficient access to space, in cooperation with U.S. industry, when sufficient U.S. commercial capabilities and services do not exist.

- **Maintain and Enhance Space-based Positioning, Navigation, and Timing Systems.** The United States must maintain its leadership in the service, provision, and use of global navigation satellite systems (GNSS).To this end, the United States shall:

  Provide continuous worldwide access, for peaceful civil uses, to the Global Positioning System (GPS) and its government-provided augmentations, free of direct user charges;
  Engage with foreign GNSS providers to encourage compatibility and interoperability, promote transparency in civil service provision, and enable market access for U.S. industry;
  Operate and maintain the GPS constellation to satisfy civil and national security needs, consistent with published performance standards and interface specifications. Foreign positioning, navigation, and timing (PNT) services may be used to augment and strengthen the resiliency of GPS; and
  Invest in domestic capabilities and support international activities to detect, mitigate, and increase resiliency to harmful interference to GPS, and identify and implement, as necessary and appropriate, redundant and back-up systems or approaches for critical infrastructure, key resources, and mission-essential functions.

- **Develop and Retain Space Professionals.** The primary goals of space professional development and retention are: achieving mission success in space operations and acquisition; stimulating innovation to improve commercial, civil, and national security space capabilities; and advancing science, exploration, and discovery. Toward these ends, departments and agencies, in cooperation with industry and academia, shall establish standards, seek to create opportunities for the current space workforce, and implement measures to develop, maintain, and retain skilled space professionals, including engineering and scientific personnel and experienced space system developers and operators, in government and commercial workforces. Departments and agencies also shall promote and expand public-private partnerships to foster educational achievement in Science, Technology, Engineering, and Mathematics (STEM) programs, supported by targeted investments in such initiatives.

- **Improve Space System Development and Procurement.** Departments and agencies shall:

  Improve timely acquisition and deployment of space systems through enhancements in estimating costs, technological risk and maturity, and industrial base capabilities;
  Reduce programmatic risk through improved management of requirements and by taking advantage of cost-effective opportunities to test high-risk components, payloads, and technologies in space or relevant environments;
  Embrace innovation to cultivate and sustain an entrepreneurial U.S. research and development environment; and
  Engage with industrial partners to improve processes and effectively manage the supply chains.

- **Strengthen Interagency Partnerships.** Departments and agencies shall improve their partnerships through cooperation, collaboration, information sharing, and/or alignment of common pursuits. Departments and agencies shall make their capabilities and expertise available to each other to strengthen our ability to achieve national goals, identify desired outcomes, leverage U.S. capabilities, and develop implementation and response strategies.

International Cooperation

Strengthen U.S. Space Leadership. Departments and agencies, in coordination with the Secretary of State, shall:

- Demonstrate U.S. leadership in space-related fora and activities to: reassure allies of U.S. commitments to collective self-defense; identify areas of mutual interest and benefit; and promote U.S. commercial space regulations and encourage interoperability with these regulations;
- Lead in the enhancement of security, stability, and responsible behavior in space;
- Facilitate new market opportunities for U.S. commercial space capabilities and services, including commercially viable

terrestrial applications that rely on government-provided space systems;

- Promote the adoption of policies internationally that facilitate full, open, and timely access to government environmental data;
- Promote appropriate cost- and risk-sharing among participating nations in international partnerships; and
- Augment U.S. capabilities by leveraging existing and planned space capabilities of allies and space partners.

Identify Areas for Potential International Cooperation. Departments and agencies shall identify potential areas for international cooperation that may include, but are not limited to: space science; space exploration, including human space flight activities; space nuclear power to support space science and exploration; space transportation; space surveillance for debris monitoring and awareness; missile warning; Earth science and observation; environmental monitoring; satellite communications; GNSS; geospatial information products and services; disaster mitigation and relief; search and rescue; use of space for maritime domain awareness; and long-term preservation of the space environment for human activity and use.

The Secretary of State, after consultation with the heads of appropriate departments and agencies, shall carry out diplomatic and public diplomacy efforts to strengthen understanding of, and support for, U.S. national space policies and programs and to encourage the foreign use of U.S. space capabilities, systems, and services.

Develop Transparency and Confidence-Building Measures. The United States will pursue bilateral and multilateral transparency and confidence-building measures to encourage responsible actions in, and the peaceful use of, space. The United States will consider proposals and concepts for arms control measures if they are equitable, effectively verifiable, and enhance the national security of the United States and its allies.

Preserving the Space Environment and the Responsible Use of Space

Preserve the Space Environment. For the purposes of minimizing debris and preserving the space environment for the responsible, peaceful, and safe use of all users, the United States shall:

- Lead the continued development and adoption of international and industry standards and policies to minimize debris, such as the United Nations Space Debris Mitigation Guidelines;
- Develop, maintain, and use space situational awareness (SSA) information from commercial, civil, and national security sources to detect, identify, and attribute actions in space that

are contrary to responsible use and the long-term sustainability of the space environment;

- Continue to follow the United States Government Orbital Debris Mitigation Standard Practices, consistent with mission requirements and cost effectiveness, in the procurement and operation of spacecraft, launch services, and the conduct of tests and experiments in space;
- Pursue research and development of technologies and techniques, through the Administrator of the National Aeronautics and Space Administration (NASA) and the Secretary of Defense, to mitigate and remove on-orbit debris, reduce hazards, and increase understanding of the current and future debris environment; and
- Require the head of the sponsoring department or agency to approve exceptions to the United States Government Orbital Debris Mitigation Standard Practices and notify the Secretary of State.

Foster the Development of Space Collision Warning Measures. The Secretary of Defense, in consultation with the Director of National Intelligence, the Administrator of NASA, and other departments and agencies, may collaborate with industry and foreign nations to: maintain and improve space object databases; pursue common international data standards and data integrity measures; and provide services and disseminate orbital tracking information to commercial and international entities, including predictions of space object conjunction.

Effective Export Policies

Consistent with the U.S. export control review, departments and agencies should seek to enhance the competitiveness of the U.S. space industrial base while also addressing national security needs.

The United States will work to stem the flow of advanced space technology to unauthorized parties. Departments and agencies are responsible for protecting against adverse technology transfer in the conduct of their programs.

The United States Government will consider the issuance of licenses for space-related exports on a case-by-case basis, pursuant to, and in accordance with, the International Traffic in Arms Regulations, the Export Administration Regulations, and other applicable laws, treaties, and regulations. Consistent with the foregoing space-related items that are determined to be generally available in the global marketplace shall be considered favorably with a view that such exports are usually in the national interests of the United States.

Sensitive or advanced spacecraft-related exports may require a government-to-government agreement or other acceptable arrangement.

National Security Space Guidelines

The Secretary of Defense and the Director of National Intelligence, in consultation with other appropriate heads of departments and agencies, shall:

- Develop, acquire, and operate space systems and supporting information systems and networks to support U.S. national security and enable defense and intelligence operations during times of peace, crisis, and conflict;
- Ensure cost-effective survivability of space capabilities, including supporting information systems and networks, commensurate with their planned use, the consequences of lost or degraded capability, the threat, and the availability of other means to perform the mission;
- Reinvigorate U.S. leadership by promoting technology development, improving industrial capacity, and maintaining a robust supplier base necessary to support our most critical national security interests;
- Develop and implement plans, procedures, techniques, and capabilities necessary to assure critical national security space-enabled missions. Options for mission assurance may include rapid restoration of space assets and leveraging allied, foreign, and/or commercial space and nonspace capabilities to help perform the mission;
- Maintain and integrate space surveillance, intelligence, and other information to develop accurate and timely SSA. SSA information shall be used to support national and homeland security, civil space agencies, particularly human space flight activities, and commercial and foreign space operations;
- Improve, develop, and demonstrate, in cooperation with relevant departments and agencies and commercial and foreign entities, the ability to rapidly detect, warn, characterize, and attribute natural and man-made disturbances to space systems of U.S. interest; and
- Develop and apply advanced technologies and capabilities that respond to changes to the threat environment.

Source: Office of Science and Technology Policy, National Space Policy of the United States of America, Washington, D.C. June 28, 2010; available from http://www.whitehouse.gov/sites/default/files/national_space_policy_6-28-10.pdf.

7

Directory of Organizations

U.S. Organizations

Advanced Research Projects Agency (ARPA) (1958–)
www.darpa.mil

The shock of the *Sputniks* was the proximate cause of the creation of ARPA, announced by Secretary of Defense Neil McElroy on November 15, 1957, and established in February 1958. ARPA was established to consolidate DoD technology development in one agency and was particularly concerned with preventing duplication of effort on missile and space programs. Roy W. Johnson, a former General Electric executive, was chosen as the first ARPA Director, and Dr. Herbert York of the PSAC was selected to serve as Chief Scientist. According to York, Secretary McElroy initially gave ARPA two assignments: control over all space programs and the task of preventing another situation like Sputnik from happening again to the United States. For a brief but intense time, ARPA directed nearly all U.S. space R&D efforts, including efforts toward human spaceflight; for example, in February 1958 it assumed control of the WS-117L program and this became the agency's single most important space project, accounting for $152 million or nearly one-third of ARPA's 1958 budget. In February 1958, President Eisenhower indicated his strong preference for keeping civil space efforts within ARPA in order to avoid duplication of effort but his opinion changed through Science Adviser James Killian's efforts to draft the legislation for NASA. In August 1958, Deputy Secretary of Defense Donald Quarles and Killian reached a compromise that directed NASA to design and build the capsules for manned spaceflight while ARPA would continue to concentrate on the boosters required for this mission. The response of other space actors toward ARPA was mixed. The Army was probably the most supportive of the new agency. It desired to reduce the potential for Air Force control of military space through a strong organization at the OSD level and believed that ARPA might make better use of Army space expertise by removing the restrictions on Army development of missiles with ranges greater than 200 miles. Killian and the PSAC were also supportive of the new agency, mainly because centralizing DoD space efforts gave the scientists more room to maneuver in crafting a civilian space agency. Congress initially viewed ARPA as inadequate. The other services and much of the aerospace industry generally were not very supportive of ARPA. The Air Force was the most opposed to the creation and operation of a strong ARPA, believing that a new strong space organization at the OSD level could derail its efforts to become predominant in space within DoD. General Schriever was among those who were most openly and consistently critical of ARPA; in congressional testimony in April 1959, he even recommended that ARPA be abolished as of July 1, 1959. Following implementation of the Defense Reorganization Act of 1958, ARPA's role in directing military space R&D began to decline. In December 1958, York was appointed as the first Director of Defense Research & Engineering (DDR&E) and was authorized to supervise all DoD research and engineering activities. ARPA's charter was limited and placed under the control of the DDR&E in the revised DoD Directive 5105.15 issued on March 17, 1959. In August 1959, York proposed to George Kistiakowsky, Eisenhower's new Special Assistant for Science and Technology, that primary responsibility for most military space R&D be returned from ARPA to the services subject to the overall supervision of the DDR&E. Eisenhower was initially skeptical of this proposal but acquiesced after it received the backing of McElroy. ARPA's name was changed to the Defense Advanced Research Projects Agency (DARPA) in 1972.

Air (and Aerospace) Defense Command (ADC)
(1946–1980)

ADC was formed in 1946 to provide for the defense of United States airspace but was not involved in space issues until the 1970s when it proposed an air-launched, direct assent ASAT known as

Project Spike. This system was to be launched from an F-106 interceptor and use a modified Standard AGM-78 antiradar missile with an explosive warhead but was never built. On January 15, 1968, ADC was renamed Aerospace Defense Command, reflecting transfer of the Program 437 ASAT system from Air Force Systems Command to ADC in November 1963. However, ADC's organizational clout declined due to the relative decrease in the Soviet bomber threat and growing power of Soviet ICBMs and submarine-launch ballistic missiles (SLBMs). ADC was deactivated as an Air Force major command on March 31, 1980.

Air Force Ballistic Missile Division (AFBMD) (1957–1961)

The AFBMD was formed in 1957 as a continuation of the Western Development Division (WDD) under the command of General Schriever. Before NASA was established, AFBMD pushed hard to gain responsibility for a major manned spaceflight program, proposing in January 1958 a comprehensive plan for manned spaceflight and launching lunar probes beginning in 1959. AFBMD released the "USAF Manned Military Space System Development Plan" (MISS) on April 25, 1958. The MISS Plan projected a Moon landing by 1965 at a cost of only $1.5 billion but was formally rejected by ARPA director Roy Johnson in July; President Eisenhower formally gave NASA primary authority over U.S. manned spaceflight efforts in August and by November 1958 this effort had evolved into Project Mercury. AFBMD work was transferred to Air Force Systems Command (AFSC) in 1961.

Air Force Space Command (AFSPC) (1982–)
www.afspc.af.mil

AFSPC was formed on September 1, 1982, to consolidate, centralize, and focus many of the Air Force's space efforts; it was the most important space-related Air Force organizational development of the Cold War era and the first completely new major command formed by the Air Force in 32 years. At its inception AFSPC was made a major space player within the Air Force but was by no means designed to be the service's sole space organization or a replacement for many of the existing space organizations such as AFSC's Space Division. AFSPC was the product of many factors and significant internal bureaucratic struggles within the Air

Force. While AFSPC would over time gain operational control over more space systems, and space activities generally would become more operationally oriented, the somewhat limited scope of AFSPC turf as specified at its creation would not change appreciably. When formed, AFSPC was the smallest Air Force major command; it consisted of only three bases and four stations, some 6,000 Air Force civilian and military personnel, and about 2,000 contractors worldwide. The first two military satellite systems AFSPC took responsibility for were the Defense Meteorological Support Program (DMSP) and the Defense Support Program (DSP) systems. By 1986, AFSPC had organized itself into a wing and squadron administrative structure similar to a traditional Air Force major command and in October 1987 the Consolidated Space Operations Center (CSOC) completed the process of taking primary responsibility for operational control of the Air Force Satellite Control Network away from the Air Force Satellite Control Facility (AFSCF) at Onizuka (Sunnyvale) AFS. By 1989, AFSPC had grown to include some 8,400 Air Force military and civilian members, 5,400 contractor personnel, and operational responsibility for 35 installations worldwide. At this time, AFSPC exercised operational control over the DSP, Global Positioning System (GPS), and DMSP satellite systems from the CSOC, and plans were well underway to transfer control of the Defense Satellite Communications System (DSCS) and Fleet Satellite Communications System (FLTSATCOM) systems as well. Despite this growth in AFSPC personnel, facilities, and operational control over space systems, it is difficult to discern exactly what type of impact this new command had on the development of military space plans and doctrine. In this regard, AFSPC fell short of the high expectations held by some who supported its creation.

Air Force Systems Command (AFSC) (1961–1992)

AFSC was created in 1961 to consolidate all Air Force research, development, and acquisition efforts. AFSC initially had a major focus on space systems, including a push from the secret Gardner Report calling for AFSC to spearhead an accelerated and very ambitious program including manned spaceflight, space weapons, reconnaissance systems, large boosters, space stations, and even a lunar landing by 1967–1970. In May 1964, AFSC sought permission to develop a nonnuclear ASAT capability based on the Thor launcher in a project known first as Program 437 Y and then as

Program 922. AFSC was characterized primarily by a technical or R&D outlook rather than an operations focus. The creation of the Space and Missile Systems Organization (SAMSO) within AFSC perpetuated the R&D mind-set within the space-development community and the general developer-user split within the national security space (NSS) mission. SAMSO reported to AFSC and was the Air Force's primary space and ballistic missile R&D organization. AFSC was merged with Air Force Logistics Command to create Air Force Material Command in 1992.

Air Research and Development Command (ARDC) (1950–1961)

ARDC was formed in 1950 to aid the Air Force's research and development efforts. Its earliest space efforts included the WS-117L satellite system initiated in March 1955, studies begun in 1956 that became the satellite inspection (SAINT) program, System Development Directive 464 in November 1957 that marked the official start of the Air Force's Dyna-Soar program, and the October 19, 1959, Project Bold Orion ASAT test. In 1961, ARDC gained responsibility for acquisition and was renamed Air Force Systems Command.

Arms Control and Disarmament Agency (ACDA) (1961–1997)

ACDA was established in 1961 as a separate part of the State Department and given responsibility for nonproliferation and disarmament strategies. ACDA representatives were active during the deliberations of the NSAM 156 Committee beginning in 1962 and were eventually successful in moving the U.S. position toward support for banning nuclear weapons from space, setting in motion the informal and formal initiatives that eventually led to the international declaratory ban on placing nuclear weapons or weapons of mass destruction in outer space in UNGA Resolution 1884 on October 17, 1963, and culminating in the Outer Space Treaty of 1967. ACDA functions were merged into the State Department in 1997.

Army Ballistic Missile Agency (ABMA) (1956–1960)

ABMA was established at the Redstone Arsenal in Huntsville, Alabama, in 1956 to develop ballistic missile technology for the Army. Dr. Wernher von Braun and his team of missile engineers that developed Germany's V-2 were brought to the Redstone Arsenal in 1950 and were the foundation for ABMA. Incoming Secretary of Defense Neil McElroy was visiting ABMA on October 4, 1957, when the news of *Sputnik* broke. Following this news flash, von Braun and ABMA Commander Major General John Medaris cornered McElroy and asserted that ABMA's Project Orbiter could place a satellite in orbit within 90 days, making good on that promise by launching *Explorer* I, the first U.S. satellite, atop a Jupiter-C (Juno) vehicle derived from the V-2 on January 31, 1958. In 1958, ARPA tasked the von Braun team at ABMA to study and design a 1.5 million pound thrust booster known as Saturn C-1, even though there was no specific military rationale for building a booster this large. In October 1958, Deputy Secretary of Defense Donald Quarles and Keith Glennan, NASA's first Administrator, tried to transfer the Army-sponsored Jet Propulsion Laboratory (JPL) at the California Institute of Technology and the von Braun team at ABMA to NASA. Medaris and Army Secretary Wilber Brucker vigorously fought against this proposal and the National Aeronautics and Space Council (NASC) had to work out a compromise allowing JPL to become part of NASA and the von Braun team to remain under ABMA but work on Saturn under contract to NASA. The continuing struggle of ABMA to remain a major player in the national space program and to retain control over the von Braun team next came to a head in the summer and fall of 1959. Science Adviser George Kistiakowsky, Quarles, DDR&E Herbert York, and eventually President Eisenhower all became convinced that ABMA should be put under NASA. Congress approved this realignment in January 1960 and the von Braun team along with most other ABMA personnel and facilities transferred to NASA's Marshall Space Flight Center on July 1, 1960.

Central Intelligence Agency (CIA) (1947–)
www.cia.gov

CIA was created by the National Security Act of 1947; its primary responsibility is to gather intelligence on and carry out covert action against actors hostile to U.S. interests. As the result of Technological Capabilities Panel (TCP) recommendations, CIA began developing sophisticated technical intelligence collection platforms in the mid-1950s. At a secret Oval Office meeting on November 24, 1954, President Eisenhower verbally authorized CIA to begin developing U-2 aircraft with Air Force support, later stipulating that only CIA pilots fly the planes. At a similar meeting on February 7, 1958, Eisenhower decided to make the CIA, rather than the Air Force, primarily responsible for the development of a reconnaissance satellite using the recoverable film method under the Corona Program. CIA was also very involved in the creation of the National Reconnaissance Office (NRO) in 1961 and continues to be an essential part of NRO operations, developing a long series of successful spysats under NRO's Program B and its successors and providing NRO's second largest personnel contingent behind the Air Force.

Department of Defense (DoD) (1947–)
www.defense.gov

DoD was created by the National Security Act of 1947 and its amendments by establishing the Department of the Air Force and a Secretary of Defense over the Air Force and former Departments of War and Navy. DoD and its components perform all NSS activities. Desires to conduct NSS activities more effectively and efficiently have contributed to changes in DoD's organizational structure including the creation of ARPA in 1958 and the DoD Reorganization Act of 1958. DoD's specific space responsibilities are listed in the National Aeronautics and Space Act of 1958 that created NASA. For much of the Cold War, DoD space activities were classified in accordance with the "Blackout" Directive issued on March 23, 1962, that prohibited advance announcement and press coverage of all military space launches, forbade use of the names of military space programs such as Midas and Samos, and replaced program names with program numbers. In January 1963, NASA Administrator James Webb and Secretary of Defense Robert McNamara signed an agreement to allow DoD experiments on some Gemini missions in a program known as Blue Gemini. Interactions between NASA and DoD in the 1970s and 1980s were important in the design and operations of the Space Transportation System (STS) or space shuttle program. When the STS ran into significant political and budgetary problems during the Carter Administration, DoD stepped in to help save the program largely due to the shuttle's projected capability to launch huge spy satellites. Following the *Challenger* disaster on January 28, 1986, DoD helped to produce a new U.S. Space Launch Strategy and the Space Launch Recovery Plan specifying the United States would rely on a balanced mix of STS and expendable launch vehicles (ELVs) with critical payloads designed for launch by either system. In the 1980s, DoD first developed a four-part military space typology used in DoD space-policy statements: space support, force enhancement, space control, and force application. In May 1987, DoD initiated a joint program to develop a new ELV program known as the Advanced Launch System (ALS). The goals of the ALS program included the development of a flexible and reliable family of modular launch vehicles which could easily be configured for specific needs and lowering the cost per pound to LEO initially to $1,500 and then to $300 by the late 1990s, but the system was never built. A second major joint DoD space launch technology development effort was emphasized by President Reagan in his 1986 State of the Union Address and was known as the X-30 or the National Aerospace Plane (NASP). The goal of the NASP program was to build an experimental manned single-stage-to-orbit vehicle which would take off like an airplane, fly into space, and then return to land like an airplane but the program proved to be extremely challenging and was abandoned in 1992.

Jet Propulsion Laboratory (JPL) (1936–)
www.jpl.nasa.gov

JPL was founded in 1936 by California Institute of Technology graduate students and received sponsorship from the U.S. Army. In December 1958, JPL became the first outside organization to become part of NASA after Deputy Secretary of Defense Donald Quarles and Keith Glennan, NASA's first Administrator, worked through the details of its transfer and overcame Army objections. JPL is the world's foremost center for developing and operating robotic probes and planetary missions including, Voyager, Viking, Galileo, Cassini-Huygens, Dawn, Mars Rovers, and the Explorer, Mariner and Ranger series.

Joint Chiefs of Staff (JCS) (1942–)
www.jcs.mil

The United States JCS was formed in 1942 as a unified high command of the Army and Navy during World War II. It is the highest military structure in DoD and is currently comprised of the Chairman, Vice Chairman, Chiefs of Staff from the Army, Navy, Air Force, and Marines, and the Joint Staff. JCS has periodically considered missile and space matters, but it is not usually the most focused or persistent actor on these issues. In March 1950, JCS recommended and Secretary of Defense Louis Johnson approved giving the Air Force exclusive jurisdiction over long-range missiles but did not define this term, contributing to considerable friction between the Army and Air Force over missile development. In September 1958, Chief of Naval Operations Arleigh Burke and the JCS proposed creation of a unified space command but this approach was opposed by the Air Force and rejected by Secretary of Defense Neil McElroy. During interagency debates in the NASM 156 deliberations in 1962, JCS initially strongly opposed banning nuclear weapons from space, arguing that a ban could not be verified and would preclude military options. Despite this opposition of the JCS, the NSAM 156 Committee and the Department of State advanced U.S. positions at the UN that led to UNGA Resolutions 1884 and 1962 in October and December 1963. Following ratification of the Outer Space Treaty In 1967, Secretary of Defense Robert McNamara did not publicly reveal that JCS regarded the Soviet fractional orbit bombardment system (FOBS) as a considerable security concern as a possible first-strike weapon able to avoid much of the U.S. early warning system by attacking from the south. During 1987, JCS established classified minimum performance requirements for a Phase I strategic defense deployment and Secretary of Defense Caspar Weinberger announced on September 18 that the Defense Acquisition Board had formally moved six parts of the Phase I SDI program past the demonstration and validation milestone of the defense acquisition process.

National Advisory Committee on Aeronautics (NACA) (1915–1958)

NACA was established in 1915 to promote aeronautical research in World War I. In 1957–1958, President Eisenhower's Science Adviser James Killian and the President's Science Advisory Committee (PSAC) felt it was important to emphasize space science and wished to avoid a military focus on this new medium; they felt a new civilian space agency modeled after and built from NACA was needed. Before NASA was created, the Air Force pushed to be in charge of a major manned space program, but Killian and NACA Director Hugh Dryden strongly opposed the Air Force's plan and urged Eisenhower to make the new civil space agency primarily responsible for the manned mission. On July 29, 1958, Dryden met with ARPA Director Roy Johnson and Secretary of Defense Neil McElroy to discuss the future management of manned space programs, but this group was unable to resolve their organizational differences on this issue. NASA inherited NACA facilities and personnel when it was created on October 1, 1958.

**National Aeronautics and Space Administration
(NASA) (1958–)**
www.nasa.gov

Under the National Aeronautics and Space Act, NASA replaced NACA on October 1, 1958, and became America's civil space agency. Creation of NASA was another response to the shock of the *Sputniks* and marked the formal beginning of two separate but closely related and imprecisely delineated American space programs. NASA's rapid creation and its broad powers reflected the desire for a new civilian space organization and primary importance of the civil space mission. At the same time, the process of creating NASA also highlighted the perceived role of space in national security and the tone of the National Aeronautics and Space Act expressed these latter considerations to greater degree than had the administration's original proposal. NASA initially lacked specific space expertise in many areas, especially booster development, but after President Eisenhower decided in August 1958 that it would be responsible for human spaceflight and President Kennedy's Moon Landing Challenge of May 1961, it grew very rapidly; NASA's budget went from $964 million in 1961 to $5.1 billion by 1964. The next significant interactions between NASA and DoD came in the 1970s and 1980s over the design and operations of the shuttle program. NASA decided to pursue a large "national" shuttle design able to accommodate the most important potential users and satisfy the military in particular. The Air Force set a number of challenging performance criteria including a requirement to lift a 65,000-pound payload and moved NASA toward a lifting body design known as the Thrust-Assisted Orbiter Shuttle (TAOS) that was formally approved as the Space Transportation System (STS) by President Nixon on January 5, 1972. Following the *Challenger* disaster on January 28, 1986, NASA produced a new U.S. Space Launch Strategy and the Space Launch Recovery Plan. In May 1987, DoD initiated a joint program with NASA to develop a new ELV program known as the Advanced Launch System (ALS). A second major joint NASA space launch technology development effort was emphasized by President Reagan in his 1986 State of the Union Address and was known as the X-30 or the National Aerospace Plane (NASP). More recent NSS-related issues with NASA have also focused mainly on launch issues. NSPD-4, "National Space Launch Strategy," issued on July 10, 1991, supported assured access through a mixed fleet of shuttles and ELVs and directed that NASA jointly undertake development of a new medium- to heavy-lift vehicle to reduce launch costs while improving reliability and responsiveness. NSTC-2 directed NASA to work on a single environmental monitoring system that would meet their requirements and save money. NSTC-3 reflected the failure of Landsat 6 in 1993, transitioned Landsat 7 from DoD to NASA. NSTC-4 was an attempt to sustain and revitalize U.S. launch systems by making NASA the lead agency for developing next-generation reusable launch vehicles. President Bush's National Space Transportation Policy issued in December 2004 made the Secretary of Defense responsible for maintaining two EELV providers, called for the Secretary to reevaluate this policy with the Director of National intelligence and NASA Administrator no later than 2010, and charged the NASA Administrator to develop options to meet potential exploration-unique requirements for heavy lift beyond the capabilities of the EELV that emphasize the potential for using EELV derivatives as well as evaluations of the comparative costs and benefits of a new dedicated heavy-lift launch vehicle or options based on the use of shuttle-derived systems. The Obama Administration did not request funding for Constellation in the NASA budget submitted to Congress in February 2010.

National Aeronautics and Space Council (NASC) (1958–1973)

NASC was created by the National Aeronautics and Space Act of 1958 to focus on developing space policy at the White House. It was the result of compromises in July 1958 that allowed a modified version of the House's Civilian-Military Liaison Committee between NASA and DoD and creation of the NASC at the White House. In its first major action in December 1958 the NASC crafted a compromise allowing JPL to become part of NASA and the von Braun team to remain under ABMA but work on Saturn under contract to NASA. NSC 5918, "U.S. Policy on Outer Space," was drafted by the NASC, discussed at the NSC meeting on December 29, 1959, and approved by President Eisenhower on January 26, 1960. The Kennedy Administration made Vice President Johnson chair of the NASC and Johnson used the council in developing Kennedy's Moon landing challenge of May 25, 1961. Just one year later, however, following a perceived lack of NASC success in fostering interagency cooperation, on May 26, 1962, Kennedy issued National Security Action Memorandum (NSAM) 156, an implicit critique of the NASC and a request for the Department of State to create a high-level coordinating body (known as the NASM 156 Committee) to address this problem. The NASC was not often used in the Johnson and Nixon Administrations, and it was disbanded in 1973.

National Geospatial-Intelligence Agency (NGA) (1996–)
www.nima.mil

In 1996 the National Imagery and Mapping Agency (NIMA) was created to expand geospatial intelligence development and consolidate these functions within one organization; in 2003 the name of NIMA was changed to NGA to reflect the revolutionary fusion of many intelligence products into geospatial intelligence. NGA works closely with NRO and commercial imagery providers to develop imagery and map-based intelligence solutions for U. S. national defense, homeland security, and safety of navigation.

National Oceanic and Atmospheric Agency (NOAA) (1970–)
www.noaa.gov

NOAA was founded in 1970 as a scientific agency within the Department of Commerce to study and monitor the oceans and atmosphere. NOAA operates the National Environmental Satellite, Data, and Information Service. NSTC-2 in May 1994 directed NOAA, DOD, and NASA to work on a single environmental monitoring system that would meet their requirements and save money. The National Polar-Orbiting Environmental Satellite System (NPOESS) program resulted from this directive; NPOESS repeatedly ran into problems in meeting its requirements on time and on budget resulting in termination of the NPOESS program and a return to separate environmental monitoring systems in President Barack Obama's FY 2011 budget request submitted to Congress in February 2010.

National Reconnaissance Office (NRO) (1961–)
www.nro.gov

NRO was established in 1961 to build and operate all U.S. intelligence collection satellites. After difficulties in developing and successfully operating the first U.S. spysats in 1958–1960, Science Adviser George Kistiakowsky and others began to suspect that problems with priorities and organizations were causing more problems than technological challenges facing these programs. In particular, Kistiakowsky believed the Air Force was putting too much effort into the electro-optical data return Samos program, that was based on technologies he believed would not mature for some time, and that this overemphasis was disrupting the entire spysat effort. By May 26, 1960, President Eisenhower had also firmly decided U.S. spysat efforts needed to be closely reviewed; he told Kistiakowsky to set up a committee alongside the PSAC to study what corrective actions might be needed. Kistiakowsky and Defense Secretary Thomas Gates decided on a study committee composed of three people: Under Secretary of the Air Force Dr. Joseph Charyk, Deputy DDR&E John Rubel, and Kistiakowsky. This group, which came to be known as the Samos Panel, reported their recommendations at the NSC meeting on August 25. The chief recommendation, the immediate creation of an organization to provide a direct chain of command from the Secretary of the Air Force to the officers in charge of each spysat project, was enthusiastically supported by Eisenhower and approved by the NSC; Eisenhower also emphasized that he did not want the new structure to result in Air Force control. Adoption of this recommendation led directly to immediate creation of an interim organization and formal creation in August 1961 of the highly classified NRO, whose very existence was an official U.S. state secret until September 1992. NRO was created as a national level organization with Air Force, CIA, and Navy participation and Charyk was chosen as the first Director, establishing the precedent that the NRO Director is usually the Under Secretary or an Assistant Secretary of the Air Force while the Deputy Director is usually from CIA. Creation of the NRO was another vote of no confidence in the ability of the Air Force to manage spy-satellite programs through more normal channels. More importantly, it ended Air Force plans for Strategic Air Command (SAC) to operate the Samos system and thus moved these most important intelligence data streams away from military operators. Overall, the creation of NRO and the extremely tight control of spysat intelligence data at the highest levels seems to have made these national overhead collection assets responsive to top decision makers within the government, but it also initiated a system whereby this most valuable of all military reconnaissance information was generally not directly available to the military, perhaps even during wartime. In the 1960s, NRO argued against developing the Manned Orbiting Laboratory; it had been leery of the idea of manned reconnaissance systems from the outset, reasoning that a manned system might present more of a provocation to the Soviets, that the contributions of manned operators in space would not be significant when balanced against the costs and requirements of life support systems, and that any accident involving MOL astronauts might set back all space-based intelligence gathering unacceptably. NRO agreed with the judgment of former Science Adviser Killian that there was a direct correlation between the amount of publicity given to U.S. spysat efforts and the probability of a Soviet response to these provocations. As NRO began developing the United States' fourth-generation photoreconnaissance satellites known as the KH-9 or "Big Bird," it also advanced the argument that the projected capabilities of the KH-9 would make MOL unnecessary. During the mid-1980s, NRO was eventually allowed to become the only U.S. government entity to build a backup launcher for its most important payloads in case of problems with the space shuttle, the Complementary ELV (CELV).

National Science and Technology Council (NSTC) (1993–)
www.whitehouse.gov/administration/eop/ostp

NSTC was formed by the Clinton Administration in 1993 as the lead organization for developing space policy in the White House. NSTC was directly involved in developing and implementing all space-policy efforts including: Convergence of U.S.-Polar Orbiting Environmental Satellite Systems (NSTC-2); Landsat Remote Sensing Strategy (NSTC-3); National Space Transportation Policy (NSTC-4); Global Positioning System Policy (NSTC-6); and the National Space Policy (PDD-49). The NSTC appeared to be less active during the George W. Bush Administration but may be poised for greater action in the Obama Administration.

National Security Council (NSC) (1947–)
www.whitehouse.gov/administration/eop/nsc

NSC was established by the National Security Act of 1947 as a White House council to address national security and foreign policy. It is comprised of the President, Vice President, and the Secretaries of State and Defense; the Chairman of the JCS and the

Director of Central Intelligence (now Director of National Intelligence) are the statutory advisers, and it has a support staff under the National Security Adviser. The NSC was directly involved in crafting the first U.S. space policy and has played a part in developing every subsequent national space policy. During the spring of 1955, the NSC Planning Board, Special Assistant to the President on Government Operations Nelson Rockefeller, and Assistant Secretary of Defense for Research & Development Donald Quarles developed the draft of NSC-5520 by reviewing and analyzing the differing goals and requirements of the TCP "freedom of space" objective, the WS-117L program, the U.S. IGY satellite proposal, and several military requirements and booster considerations. The subsequent role of the NSC in developing space policy has varied by administration and changes in other White House space-policy-making structures.

National Space Council (NSpC) (1989–1993)

The NSpC in the **George** H. W. Bush Administration was a recreation of a high-level space-policy coordination mechanism similar to the NASC that operated at the White House from 1958 to 1973. NSpC was chaired by Vice President Dan Quayle and included the Secretaries of State, Treasury, Defense, Commerce, and Transportation, Director of the Office of Management and Budget, White House Chief of Staff, National Security Adviser, Director of OSTP, Director of Central Intelligence, and NASA Administrator; there was also an executive secretary and small support staff. The NSpC was directly involved in developing and implementing all space policy during the Bush Administration but was criticized for developing space-centric approaches that were divorced from political realities; it was disbanded in 1993 and the National Science and Technology Council assumed its responsibilities.

Naval Research Laboratory (NRL) (1923–)
www.nrl.navy.mil

NRL was founded in 1923 under the direction of Thomas Edison; its primary purpose is to improve naval technology. In 1955, NRL competed against proposals from WDD and ABMA for the honor of having its booster launch the U.S. International Geophysical Year (IGY) satellite. The NRL proposal called for the development of an upgraded version of the Navy's Viking sounding rocket capable of launching a very small satellite; on August 3 the Stewart

Committee voted 3–2 in favor of NRL's proposal and this marked the beginning of Project Vanguard. Many factors were at work influencing the close vote, including political sensitivities about legitimizing satellite overflight and a desire to avoid having a Nazi-tainted booster lead the United States into the space age. Because the NRL booster was not directly associated with any major military missile program it was better suited to maintain a more civilian face on the IGY effort. Selection of the unproven NRL booster reemphasized the civilian face of U.S. space efforts, covertly supported overflight considerations, but also set the stage for America's second-place finish in the first space race. The first U.S. attempt to launch a satellite on December 6, 1957, resulted in a televised spectacular failure when Vanguard TV-3 suffered a launch-pad explosion. NRL holds the distinction of successfully operating the first intelligence-collection satellite: the Galactic Radiation and Background (GRAB) satellite was launched on June 22, 1960, and began returning electronic intelligence data in July 1960, but this mission was not declassified until 1998. In the 1960s, NRL built the Timation satellites that were the precursor to the Global Positioning System in use today.

Office of Science and Technology Policy (OSTP) (1976–)

OSTP was established by Congress in May 1976 to create a White House organization for providing scientific information and advice to the President. It has been the leading actor in developing civil space policy and works with other White House organizations, including the National Science and Technology Council and the National Security Council, in developing national space policy.

Office of the Secretary of Defense (OSD) (1947–)
www.defense.gov/osd

OSD is the staff supporting the Secretary of Defense; it provides the civilian control over DoD by exercising policy development, planning, resource management, fiscal, and program-evaluation responsibilities. The organizational structure of OSD has varied considerably over time and the locus of control over NSS has not remained consistent but is often the most important factor in shaping NSS policies and activities. OSD's organizational agenda was a major factor in injecting more of a national security focus into the final NASA bill in 1958. In the late 1950s and early 1960s, OSD reined in unrealistic military space ambitions by rejecting a unified space command, consolidating space R&D in DoD Directive 5160.32, and taking a cautious, "building block" approach toward new military space programs. Today, the major NSS actors in OSD include the Deputy Secretary, the Under Secretaries for Policy; Acquisition, Technology and Logistics (AT&L); and Intelligence; and the Assistant Secretary for Networks & Information Integration (NII).

President's Science Advisory Committee (PSAC) (1957–1972)

PSAC was established in 1957 to upgrade the Science Advisory Council and provide a more direct means of providing expert scientific opinions and analysis to the President. PSAC had spent the last months of 1957 in a series of debates over the relative value of various potential space missions and had considered many different ways to organize the government to conduct space activity. By the end of December, a consensus emerged that scientifically oriented civil space missions ought to be the nation's top space priority and that a civilian space agency built from and modeled after NACA would be the best approach. On February 3, 1958, PSAC was formally tasked with studying space mission priorities and recommending possible organizational structures. The next day, this PSAC study, which came to be known as the Purcell Report after its chairman Edward Purcell of Harvard, was initiated and publicly announced. The most significant aspect of the Purcell Report was its de-emphasis of military space applications. PSAC also dominated much of the process of drafting the legislation for NASA. Prior service on the PSAC gave Herbert York a solid foundation for evaluating missile and space programs and the confidence to propose sweeping reorganizations of America's space programs when he served as the first DDR&E. In October 1963, the PSAC compared the relative military utility of the Gemini, X-20, and MOL programs and judged that the X-20 held the least potential. President Nixon eliminated PSAC in 1972 and its functions are now performed by the Office of Science and Technology Policy (OSTP), established in 1976.

Research and Development Corporation (RAND) (1946–)
www.rand.org

In March 1946, Commander of the U.S. Army Air Force, General Henry H. (Hap) Arnold, authorized creation of a joint project with the Douglas Aircraft Company on research and development

(RAND) that became the basis for one of the most influential think tanks in the United States. Its first report, "Preliminary Design of an Experimental World-Circling Spaceship," completed in April 1946, not only explained space physics and spaceship technical designs but also introduced almost every major military space mission that would be developed in the coming years including communications, attack assessment, weather reconnaissance, and strategic reconnaissance. RAND also became the first organization to analyze the broader political implications of opening the space age in an October 1950 report highlighting the likely psychological impression the first satellite would leave on the public and raising the issues of "overflight" and "freedom of space" by asking how the Soviets would respond to satellites flying over and photographing their territory. This report suggested that one way to test the issue of freedom of space would be first to launch an experimental U.S. satellite in an equatorial orbit that would not cross Soviet territory before attempting any satellite reconnaissance overhead the Soviet Union. Several other satellite application studies were conducted by RAND during the late 1940s and early 1950s but were hampered by very limited funding and often marginalized by the mind-set of many influential military and scientific leaders who relegated such notions to the realm of science fiction.

**Senior Interagency Group for Space
(SIG (Space)) (1981–1989)**

SIG (Space) was formed by the Reagan Administration in 1981 as a high-level space-policy coordination mechanism. Reagan's first major space policy was contained in National Security Decision Directive (NSDD)-42 and publicly announced by the President on July 4, 1982. NSDD-42 was crafted in late 1981 and the first half of 1982 by SIG (Space) chaired by Science Adviser George Keyworth. During the first half of 1982, SIG (Space) was often the scene of intense debates over the proper focus of future U.S. space efforts. Generally, Keyworth, DoD, and CIA remained unconvinced of the utility of the large-scale, permanently manned space station that NASA Administrator James Beggs and NASA Deputy Administrator Hans Mark were pushing as NASA's next major goal. Keyworth and his supporters felt that NASA should first concentrate on getting the STS to live up to its many promises before diverting its attention to the next space spectacular.

Space and Missile Systems Organization (SAMSO) (1967–1979)

On July 1, 1967, the Air Force created SAMSO by combining Space Systems Division and Ballistic Systems Division. SAMSO reported to AFSC and served as the Air Force's primary space and ballistic missile R&D organization. SAMSO had two primary subordinate organizations: the Space and Missile Test Center (SAMTEC), responsible for launching satellites and testing ballistic missiles from VAFB and the Eastern Test Range (ETR) at Cape Canaveral; and the Air Force Satellite Control Facility (AFSCF) at Sunnyvale Air Force Station (AFS), California, responsible for controlling military satellites on orbit. Efforts on the part of SAMSO to develop new ASAT capabilities to replace Program 437 were not approved within the Air Force and no major funding was sought for actual development of new systems. On October 1, 1979, SAMSO was deactivated and its responsibilities were again divided between AFSC's newly recreated Space Division and Ballistic Missile Office.

Space Task Group (STG) (1969)

Shortly after entering office, President Richard Nixon established the STG to complete a comprehensive review of the future plans of the U.S. space program. The members of the STG were Vice President Spiro Agnew, Acting NASA Administrator Thomas Paine, Secretary of Defense Melvin Laird, and Science Adviser Lee DuBridge. On September 15, 1969, STG presented Nixon with three options for post-Apollo U.S. civil space plans. Option one called for a manned mission to Mars by 1985 supported by a 50-person space station in Earth orbit, a smaller space station in lunar orbit, a lunar base, a space shuttle to service the Earth space station, and a space tug to service the lunar stations. Option two consisted of all of the above except for the lunar projects and delayed the Mars landing until 1986. Option three included only the space station and the space shuttle, deferring the decision on a Mars mission but keeping it as a goal to be realized before the end of the century. The report estimated that annual costs for option one would be $10 billion, option two $8 billion, and option three $5 billion. Considering that NASA's budget had peaked at the height of the Moon race in 1965 at a little more than $5 billion and that political support for space spectaculars was rapidly eroding, STG recommendations seemed fiscally irresponsible and politically naive.

On the basis of the STG report and the recommendations from other major space studies during this period, Nixon formalized U.S. post-Apollo space-policy goals in March 1970 by endorsing development of a shuttle and leaving a space station or Mars mission contingent upon successful completion of a shuttle program. In March 1969, STG chairman Agnew had directed that a joint DOD-NASA study on a shuttle system be completed to support the overall STG effort. In June, DoD and NASA submitted to the STG their coordinated report on the Space Transportation System (STS), which strongly backed the development of a shuttle. By contrast, the Morrow report, also prepared for the STG, questioned the technical feasibility of a shuttle and specifically refuted the projected STS launch rates and cost estimates. Nixon formally approved development of the STS on January 5, 1972.

Strategic Defense Initiative Office (SDIO) (1984–)

SDIO was established in 1984 within the Department of Defense to help carry out President Reagan's Strategic Defense Initiative. The Future Security Strategy Study and Fletcher Study provided strong support for the authorization in NSDD-119 on January 6, 1984, to begin SDI and establish the SDIO reporting to the Secretary of Defense later that month. On April 15, Lieutenant General James Abrahamson moved from his position as Associate Administrator for Manned Spaceflight at NASA to become the first director of SDIO. His new office first demonstrated the potential of emerging defense technologies on June 10 when a kinetic energy interceptor known as the Homing Overlay Experiment launched from Meck Island in the Pacific Test Range successfully intercepted a test reentry vehicle launched atop a Minuteman ICBM from Vandenberg AFB. A series of SDIO experiments conducted on September 5, 1986, known as the Delta 180 test confirmed the ability of space-based infrared sensors and kinetic weapons to perform simulated boost-phase intercepts. While SDIO clearly advanced many missile-defense efforts, the joint composition of the SDIO staff as well as its unclear bureaucratic standing and long-term prospects also served to discourage extensive and effective cross-fertilization between the services and SDIO. In addition, because there was little attention at SDIO on developing any doctrine or concepts of operations for space weaponry and because of its research-only charter focusing solely on ballistic missile defenses, the office contributed only marginally to thinking about broader issues such as space force employment, space control, and other potential military applications of space capabilities. The name of SDIO was modified to the Ballistic Missile Defense Organization (BMDO) in 1993 and again changed in 2002 to its current name, the Missile Defense Agency (MDA).

Strategic Missiles Evaluation Committee (SMEC) (1953–1954)

SMEC was an 11-person panel established in 1953 to assess U.S. missile programs. It was established by Trevor Gardner, special assistant to secretary of the Air Force for research and development issues, and chaired by Dr. John von Neumann of Princeton University. The SMEC report of February 1954 strongly recommended the Atlas ICBM program be accelerated, called for a new management philosophy and organizational structure, and was the single most important factor in setting the structure and pace of the Atlas ICBM program from 1954 onward.

Technological Capabilities Panel (TCP) (1954–1955)

President Eisenhower's approach toward space and strategic issues was strongly influenced by the secret TCP he commissioned in March 1954. Eisenhower chose Dr. James Killian, President of the Massachusetts Institute of Technology (MIT), as chairman of the TCP and made it clear that he wanted the best minds in the country to focus on the technological problem of preventing another Pearl Harbor. As part of the TCP process, Killian and Edwin (Din) Land, founder of the Polaroid Corporation and chairman of the intelligence subcommittee of the TCP, were briefed on a wide range of potential collection methods and systems including satellites, but they became most enthused about attempting high-altitude reconnaissance overflights of the Soviet Union via a jet-powered glider they saw on the drawing boards at Clarence (Kelly) Johnson's Lockheed "Skunk Works" in Burbank, California. They recommended production of this new aircraft during a series of briefings culminating in a secret Oval Office meeting on November 24, 1954 attended by the President, Secretaries of State and Defense, as well as top Department of Defense (DoD) and Central Intelligence Agency (CIA) officials. The TCP completed a secret two-volume report and briefed the National Security Council (NSC) in February 1955. The TCP recommended several major programs including: development of IRBMs suitable for land or sea launch (leading to the Jupiter and Thor IRBMs as well as the Polaris sea-launched ballistic missile or SLBM), construction of a distant early warning (DEW) radar network to warn of a Soviet bomber attack, efforts to harden Strategic Air Command (SAC) facilities and aircraft to nuclear attack, a research program to investigate possibilities for ballistic missile defense (BMD), and continued development of the Atlas ICBM as rapidly as possible.

United States Air Force (USAF) (1947–)
www.usaf.mil

The Air Force was created from the Army Air Forces as a separate service by the National Security Act of 1947. It has been and remains the military department responsible for most NSS activity, generally accounting for more than 80 percent of the personnel and budget supporting NSS. Through RAND the Air Force was involved in the very earliest U.S. thinking about space and in March 1955 it initiated the WS-117L satellite system, the first U.S. space program. Due to his space-for-peaceful-purposes policy President Eisenhower repeatedly stifled Air Force space ambitions through a series of decisions, including making CIA responsible for the Corona spysat program, making NASA responsible for human spaceflight, and creating NRO as a joint Air Force–CIA–Navy organization. In 1961, most Air Force space activity was consolidated into the new Air Force Systems Command and in 1982 was again realigned under Air Force Space Command (AFSPC), the command that remains responsible for most Air Force space activity today.

United States Space Command
(USSPACECOM) (1985–2002)

USSPACECOM was established on September 23, 1985, as a unified command to bring together all branches of the military in using space capabilities most effectively. Establishment of USSPACECOM was a highly significant development within DoD's organizational structure for space and involved considerably more controversy than did the creation of AFSPC. In 1983 and 1984 the other services, and the Navy in particular, needed to be convinced of the need for a unified space command. USSPACECOM became the smallest of a group of only nine unified commands that control all combat forces within the U.S. military. USSPACECOM was placed above AFSPC and contained the Naval Space Command

and the Army Space Office as well. The responsibilities for USS-PACECOM were the same as those given to AFSPC with the addition of setting requirements for and planning the operational ballistic missile-defense system that might result from the SDI program. As a unified command with broad responsibilities and representing all services, USSPACECOM had slightly more of a space-user outlook rather than the space-operator outlook predominant in AFSPC. The position of Commander in Chief (CINC) of USSPACECOM was created as a four-star billet and originally given triple-hatted responsibilities as Commander of North American Aerospace Defense Command (CINCNORAD), US-CINCSPACE, and AFSPC Commander. During the existence of USSPACECOM, the Commander was always from the Air Force while the Deputy Commander was drawn from the Navy. This unified command was the first military organization specifically charged with preparing for the space-combat mission, and it provided strong advocacy for military space systems during the last few years of the Cold War. USSPACECOM proved to be a fairly influential military organization for advocating military space forces and missions to support the space-control and high-ground schools. The doctrinal impact of USSPACECOM was also quite significant, although less pronounced than its general advocacy role might suggest, and it is not clear how much impact this advocacy had on overall military space plans, programs, and doctrine. Most of USSPACECOM's efforts were generally focused on near-term attempts to push or save specific ongoing programs such as SDI or the MHV ASAT, leading to less effort in developing the type of long-range military space requirements and doctrine that might help to support future military space systems. On October 2, 2002, USSPACECOM was merged into United States Strategic Command (USSTRATCOM). Although originally described as a joining of equals, in practice this major organizational shift quickly amounted to the absorption of USSPACECOM into USSTRAT-COM and left very few vestiges of the original USSPACECOM. Instead of space being the sole focus of one of just nine unified commands, under the new structure space now competes for attention among a very wide array of disparate mission areas that include deterring attacks on U.S. vital interests, ensuring freedom of action in space and cyberspace, delivering integrated kinetic and nonkinetic effects to include nuclear and information operations in support of U.S. Joint Force Commander operations, synchronizing missile-defense plans and operations, and combating weapons of mass destruction.

**United States Strategic Command
(USSTRATCOM) (1992–)**
www.stratcom.mil

USSTRATCOM was established in 1992 by President George H. W. Bush. USSTRATCOM unified strategic planning and deployment under one commander and reflected a broader outlook than the nuclear focus of Strategic Air Command during the Cold War. On October 1, 2002, USSPACECOM was merged into United States Strategic Command (USSTRATCOM); although originally described as a joining of equals, in practice this was a major organizational shift that left very few traces of the original USSPACECOM. USSTRATCOM is responsible for a wide range of important missions, including deterring attacks on U.S. vital interests, ensuring freedom of action in space and cyberspace, delivering integrated kinetic and nonkinetic effects to include nuclear and information operations in support of U.S. Joint Force Commander operations, synchronizing missile-defense plans and operations, and combating weapons of mass destruction.

Western Development Division (WDD) (1954–1957)

WDD was created by ARDC in 1954 to speed development of the Atlas ICBM. Brigadier General Bernard Schriever took command of WDD in El Segundo, California, in August 1954. Schriever was enthusiastic about developing WS-117L satellites but was primarily charged with developing the higher-priority Atlas and received little funding or support for developing satellites. He pioneered new methods of systems management that moved beyond the traditional Air Force contractor model and also explored concurrent development. Specifically, Trevor Gardner and Schriever largely were able to shield WDD from the normal ARDC financial and system review channels and, more importantly, they moved the overall systems engineering responsibility for Atlas from Convair to the newly created Ramo-Woolridge Corporation. WDD was replaced by the Air Force Ballistic Missile Division in 1957.

8

Resources

Books, Chapters in Books, and Monographs

Allison, Graham T. *Essence of Decision: Explaining the Cuban Missile Crisis.* Boston, MA: Little, Brown & Co., 1971.

Armacost, Michael H. *The Politics of Weapons Innovation: The Thor-Jupiter Controversy.* New York: Columbia University Press, 1969.

Arnold, David Christopher. *Spying from Space: Constructing America's Satellite Command and Control Systems.* College Station: Texas A&M University Press, 2005.

Ball, Desmond. *A Base for Debate: The U.S. Satellite Station at Nurrungar.* Sidney: Allen and Unwin, 1987.

Ball, Desmond. *Pine Gap: Australia and the US Geostationary Signals Intelligence Satellite Program.* Sidney: Allen and Unwin, 1988.

Ball, Desmond. *Politics and Force Levels: The Strategic Missile Program of the Kennedy Administration.* Berkeley: University of California Press, 1980.

Bamford, V. James. *The Puzzle Palace: A Report on America's Most Secret Agency.* Boston, MA: Houghton Mifflin, 1982.

Barker, Kenneth W. *Airborne and Space-Based Lasers: An Analysis of Technological and Operational Compatibility.* Maxwell AFB, AL: Center for Strategy and Technology, 1999.

Baucom, Donald R. *The Origins of SDI, 1944–1983*. Lawrence: University Press of Kansas, 1992.

Beard, Edmund. *Developing the ICBM: A Study in Bureaucratic Politics*. New York: Columbia University Press, 1976.

Beschloss, Michael R. *Mayday: Eisenhower, Khrushchev and the U-2 Affair*. New York: Harper & Row, 1986.

Blair, Bruce G. *Strategic Command and Control: Redefining the Nuclear Threat*. Washington, DC: Brookings Institution, 1985.

Boffey, Philip M., William J. Broad, Leslie H. Gelb, Charles Mohr, and Holcomb B. Noble. *Claiming the Heavens: The New York Times Complete Guide to the Star Wars Debate*. New York: Times Books, 1988.

Bottome, Edgar M. *The Missile Gap: A Study in the Formulation of Military and Political Policy*. Rutherford, NJ: Fairleigh Dickinson University Press, 1971.

Bruce-Briggs, B. *The Shield of Faith: Strategic Defense from Zeppelins to Star Wars*. New York: Touchstone Books, Simon & Schuster, 1988.

Bulkeley, Rip. *The Sputniks Crisis and Early United States Space Policy: A Critique of the Historiography of Space*. Bloomington: Indiana University Press, 1991.

Burrows, William E. *Deep Black: Space Espionage and National Security*. New York: Berkley Books, 1986.

Canan, James. *War in Space*. New York: Berkley Books, 1982.

Carter, Ashton B., and David N. Schwartz, eds. *Ballistic Missile Defense*. Washington, DC: Brookings Institution, 1984.

Chun, Clayton K.S. (Lt Col, USAF). *Shooting Down a "Star"— Program 437, the US Nuclear ASAT System and Present-Day Copycat Killers*. CADRE Paper No. 6. Maxwell AFB, AL: Air University Press, April 2000.

Codevilla, Angelo M. "Space, Intelligence and Deception." In *Soviet Strategic Deception*, eds. Brian D. Dailey and Patrick J. Parker, 467–486. Lexington, MA: Lexington Books, 1987.

Codevilla, Angelo M. *While Others Build: The Commonsense Approach to the Strategic Defense Initiative.* New York: Free Press, 1988.

Collins, John M. *Military Space Forces: The Next 50 Years.* Washington, DC: Pergamon-Brassey's, 1989.

Davies, Merton E., and William R. Harris. *RAND's Role in the Evolution of Balloon and Satellite Observation Systems and Related Space Technology.* Santa Monica, CA: RAND Corporation, 1988.

Day, Dwayne A. "Invitation to Struggle: The History of Civilian-Military Relations in Space." In *Exploring the Unknown: Selected Documents in the History of the U.S. Civil Space Program,* Vol. II: External Relationships, ed. John M. Logsdon, 233–270. Washington, DC: GPO, 1996.

Day, Dwayne A. et al., eds. *Eye in the Sky: The Story of the Corona Spy Satellites.* Washington, DC: Smithsonian Institution Press, 1999.

DeBlois, Bruce M., ed. *Beyond the Paths of Heaven: The Emergence of Space Power Thought.* Maxwell AFB, AL: Air University Press, 1999.

DeSutter, Robert Joseph. "Arms Control Verification: 'Bridge' Theories and the Politics of Expediency." PhD diss., University of Southern California, 1983.

DeVorkin, David H. *Science With a Vengeance: How the Military Created the US Space Sciences After World War II.* New York: Springer-Verlag, 1992.

Divine, Robert A. *The Sputnik Challenge.* New York: Oxford University Press, 1993.

Dolman, Everett C. *Astropolitik: Classical Geopolitics in the Space Age.* London: Frank Cass, 2002.

Eisenhower, Dwight D. *The White House Years: Waging the Peace, 1956–1961.* New York: Doubleday, 1965.

Emme, Eugene M. *The Impact of Air Power: National Security and World Politics.* Princeton, NJ: D. Van Nostrand, 1959.

Emme, Eugene M. "Presidents and Space." In *Between Sputnik and the Shuttle: New Perspectives on American Astronautics*, ed. Frederick C. Durant, III, 5–138. San Diego, CA: American Astronautical Society, 1981.

Erickson, Mark. *Into the Unknown Together: The DOD, NASA, and Early Spaceflight*. Maxwell AFB, AL: Air University Press, 2005.

Feyock, Stephanie, ed. *Presidential Decisions: National Security Council Documents*. Washington, DC: Marshall Institute, March 2006.

Freedman, Lawrence. *The Evolution of Nuclear Strategy*. New York: St. Martin's Press, 1983.

Freedman, Lawrence. *US Intelligence and the Soviet Strategic Threat*. London: Macmillan Press, 1977.

Friedman, Norman. *Seapower and Space: From the Dawn of the Missile Age to Net-Centric Warfare*. London: Chatham Publishing, 2000.

Futrell, Robert Frank. *Ideas, Concepts, Doctrine: Basic Thinking in the United States Air Force, 1961–1984*, Vol. II, Maxwell AFB, AL: Air University Press, December 1989.

Futrell, Robert Frank. *Ideas, Concepts, Doctrine: A History of Basic Thinking in the United States Air Force, 1907–1964*. Maxwell AFB, AL: Air University Press, 1971; reprint, New York: Arno Press, 1980.

Gavin, Lieutenant General James M., USA, (Ret.). *War and Peace in the Space Age*. New York: Harper & Brothers, 1958.

Giffen, Colonel Robert B., USAF. *US Space System Survivability: Strategic Alternatives for the 1990s*. Washington, DC: National Defense University Press, 1982.

Glasstone, Samuel, and Philip J. Dolan. *The Effects of Nuclear Weapons*. 3rd ed. Washington, DC: Department of Defense and Department of Energy, 1977.

Graham, Lieutenant General Daniel O., USA, (Ret.). *High Frontier: A New National Strategy*. Washington, DC: High Frontier, 1982.

Graham, Lieutenant General Daniel O., USA, (Ret.). *Shall America Be Defended? SALT II and Beyond.* New Rochelle, NY: Arlington House, 1979.

Gray, Colin S. *American Military Space Policy: Information Systems, Weapons Systems and Arms Control.* Cambridge, MA: Abt Books, 1982.

Green, Constance Mclaughlin, and Milton Lomask. *Vanguard: A History.* Washington, DC: Smithsonian Institution Press, 1971.

Gruntman, Mike. *Blazing the Trail: The Early History of Spacecraft and Rocketry.* Reston, VA: AIAA, 2004.

Hall, R. Cargill. "Clandestine Victory: Dwight D. Eisenhower and Overhead Reconnaissance in the Cold War." In *Forging the Shield: Eisenhower and National Security for the 21st Century,* ed. Dennis Showalter. Chicago, IL: Imprint Publications, 2005.

Hall, R. Cargill. "Earth Satellites: A First Look by the United States Navy." In *History of Rocketry and Astronautics: Proceedings of the Third through the Sixth History Symposia of the International Academy of Astronautics,* 253–278. San Diego, CA: Univelt, 1986.

Hall, R. Cargill. "The Origins of U.S. Space Policy: Eisenhower, Open Skies, and Freedom of Space." In *Exploring the Unknown: Selected Documents in the History of the U.S. Civil Space Program,* Vol. I: Organizing for Exploration, ed. John M. Logsdon, 213–229. Washington, DC: GPO, 1995.

Hall, R. Cargill, and Jacob Neufeld, eds. *The U.S. Air Force in Space 1945 to the 21st Century.* Washington, DC: USAF History and Museums Program, 1998.

Hallion, Richard P., ed. *The Hypersonic Revolution: Case Studies in the History of Hypersonic Technology.* 3 Vols. Bolling AFB, DC: Air Force History and Museums Program, 1998.

Handberg, Roger. *Seeking New World Vistas: The Militarization of Space.* Westport, CT: Praeger, 2000.

Harris, William R. "Counterintelligence Jurisdiction and the Double-Cross System by National Technical Means." In *Intelligence Requirements for the 1980s: Counterintelligence,* ed. Roy

Godson, 53–82. Washington, DC: National Strategy Information Center, 1980.

Harris, William R. "Soviet Maskirovka and Arms Control Verification." In *Soviet Strategic Deception*, eds. Brian D. Dailey and Patrick J. Parker, 185–224. Lexington, MA: Lexington Books, 1987.

Hays, Peter L. *United States Military Space: Into the 21st Century*. Maxwell AFB, AL: Air University Press, September 2002.

Hays, Peter L., James M. Smith, Alan R. Van Tassel, and Guy M. Walsh, eds. *Spacepower for a New Millennium: Space and U.S. National Security*. New York: McGraw-Hill, 2000.

Houchin, Roy. *US Hypersonic Research and Development: The Rise and Fall of Dyna-Soar, 1944–1963*. New York: Routledge, 2006.

Jastrow, Robert. *How to Make Nuclear Weapons Obsolete*. Boston, MA: Little, Brown & Co, 1983.

Jenkins, Dennis R. *Space Shuttle: The History of Developing the National Space Transportation System*. Marceline, MO: Walsworth Publishing, 1992.

Johnson, Dana Joyce. "The Evolution of U.S. Military Space Doctrine: Precedents, Prospects, and Challenges." PhD diss., University of Southern California, December 1987.

Johnson, Lyndon B. *The Vantage Point: Perspectives of the Presidency, 1963–1969*. New York: Holt, Rinehart and Winston, 1971.

Johnson-Freese, Joan. *Space as a Strategic Asset*. New York: Columbia University Press, 2007.

Kaplan, Fred. *The Wizards of Armageddon*. New York: Touchstone Books, Simon & Schuster, 1983.

Karas, Thomas. *The New High Ground: Systems and Weapons of Space Age War*. New York: Simon & Schuster, 1983.

Killian, James R. Jr. *Sputnik, Scientists, and Eisenhower: A Memoir of the First Special Assistant to the President for Science and Technology*. Cambridge, MA: MIT Press, 1977.

Kistiakowsky, George. *A Scientist at the White House: The Private Diary of President Eisenhower's Special Assistant for Science and Technology*. Cambridge, MA: Harvard University Press, 1976.

Klein, John J. *Space Warfare: Strategy, Principles and Policy.* New York: Routledge, 2006.

Lambakis, Steven. *On the Edge of Earth: The Future of American Space Power.* Lexington: University Press of Kentucky, 2001.

Lambeth, Benjamin S. *Mastering the Ultimate High Ground: Next Steps in the Military Uses of Space.* Santa Monica, CA: RAND, 2003.

Lambright, W. Henry, ed. *Space Policy in the 21st Century.* Baltimore, MD: Johns Hopkins University Press, 2003.

Launius, Roger D., John M. Logsdon, and Robert W. Smith, eds. *Reconsidering Sputnik: Forty Years Since the Soviet Satellite.* Amsterdam: Harwood Academic Publishers, 2000.

Levine, Alan J. *The Missile and Space Race.* Westport, CT: Praeger Publishers, 1994.

Lewis, Jonathan E. *Spy Capitalism: ITEK and the CIA.* New Haven, CT: Yale University Press, 2002.

Lindgren, David T. *Trust but Verify: Imagery Analysis in the Cold War.* Annapolis, MD: Naval Institute Press, 2000.

Lindsey, Robert. *The Falcon and the Snowman.* New York: Pocket Books, 1979.

Logsdon, John M. *The Decision to Go to the Moon: Project Apollo and the National Interest.* Cambridge, MA: MIT Press, 1970.

Lonnquest, John C., and David F. Winkler. *To Defend and Deter: The Legacy of the United States Cold War Missile Program.* Champaign, IL: U.S. Army Construction Engineering Research Laboratories, 1996.

Luongo, Kenneth N., and W. Thomas Wander, eds. *The Search for Security in Space.* Ithaca, NY: Cornell University Press, 1989.

Lupton, Lieutenant Colonel David E., USAF, (Ret.). *On Space Warfare: A Space Power Doctrine.* Maxwell Air Force Base, AL: Air University Press, June 1988.

Manno, Jack. *Arming the Heavens: The Hidden Military Agenda for Space, 1945–1995.* New York: Dodd, Mead & Co., 1984.

Mark, Hans. *The Space Station: A Personal Journey.* Durham, NC: Duke University Press, 1987.

Martel, William C., ed. *The Technological Arsenal: Emerging Defense Capabilities.* Washington, DC: Smithsonian Institution Press, 2001.

McDougall, Walter A. . . . *the Heavens and the Earth: A Political History of the Space Age.* New York: Basic Books, 1985.

McLucas, John L., with Kenneth J. Alnwick and Lawrence R. Benson. *Reflections of a Technocrat: Managing Defense, Air, and Space Programs During the Cold War.* Maxwell AFB, AL: Air University Press, 2006.

Medaris, Major General John B., USA, (Ret.). *Countdown for Decision.* New York: G. P. Putnam's Sons, 1960.

Michaud, Michael A. G. *Reaching for the High Frontier: The American Pro-Space Movement, 1972–84.* New York: Praeger, 1986.

Miller, Steven E., and Stephen Van Evera, eds. *The Star Wars Controversy.* Princeton, NJ: Princeton University Press, 1986.

Moltz, James Clay. *The Politics of Space Security: Strategic Restraint and the Pursuit of National Interests.* Palo Alto, CA: Stanford University Press, 2008.

Moore, Mike. *Twilight War: The Folly of U.S. Space Dominance.* Oakland, CA: The Independent Institute, 2008.

Mowthorpe, Matthew. *The Militarization and Weaponization of Space.* Lanham, MD: Lexington Books, 2003.

Nitze, Paul H. *From Hiroshima to Glasnost: At the Center of Decision—A Memoir.* New York: Grove Weidenfeld, 1989.

Nye, Joseph S. Jr., and James A. Schear, eds. *Seeking Stability in Space: Anti-Satellite Weapons and the Evolving Space Regime.* Lanham, MD: University Press of America and Aspen Strategy Group, 1987.

Oberg, James E. *Red Star in Orbit.* New York: Random House, 1981.

Oberg, James E. *Space Power Theory.* Peterson AFB, CO: United States Space Command, 1999.

O'Neill, Gerard K. *The High Frontier: Human Colonies in Space.* New York: William Morrow, 1977.

Ordway, Frederick I., and Mitchell R. Sharpe. *The Rocket Team.* New York: Thomas Y. Crowell, 1979.

Orman, Stanley. *Faith in G.O.D.S.: Stability in the Nuclear Age.* London: Brassey's, 1991.

Paret, Peter, ed. *The Makers of Modern Strategy: From Machiavelli to the Nuclear Age.* Princeton, NJ: Princeton University Press, 1986.

Payne, Keith B., ed. *Laser Weapons in Space: Policy and Doctrine.* Boulder, CO: Westview Press, 1983.

Payne, Keith B. *Strategic Defense: "Star Wars" in Perspective.* Lanham, MD: Hamilton Press, 1986.

Pedlow, Gregory W., and Donald E. Welzenbach. *The CIA and the U-2 Program, 1954–1974.* Washington, DC: History Staff, Center for the Study of Intelligence, Central Intelligence Agency, 1998.

Peebles, Curtis. *Battle for Space.* New York: Beaufort Books, 1983.

Peebles, Curtis. *High Frontier: The United States Air Force and the Military Space Program.* Washington, DC: Air Force History and Museums Program, 1997.

Prados, John. *The Soviet Estimate: U.S. Intelligence Analysis & Russian Military Strength.* New York: Dial Press, 1982.

Pratt, Erik K. *Selling Strategic Defense: Interests, Ideologies, and the Arms Race.* Boulder, CO: Lynne Rienner Publishers, 1990.

Preston, Bob, Dana J. Johnson, Sean Edwards, Michael Miller, and Calvin Shipbaugh. *Space Weapons, Earth Wars.* Santa Monica, CA: RAND Corporation, 2002.

Ra'anan, Uri, and Robert L. Pfaltzgraff Jr., eds. *International Security Dimensions of Space.* Hamden, CT: Archon Books, 1984.

Reiss, Edward. *The Strategic Defense Initiative.* Cambridge, UK: Cambridge University Press, 1992.

Reynolds, Glenn H., and Robert P. Merges. *Outer Space: Problems of Law and Policy.* Boulder, CO: Westview Press, 1989.

Richelson, Jeffrey T. *America's Secret Eyes in Space: The U.S. Keyhole Spy Satellite Program.* New York: Harper & Row, 1990.

Richelson, Jeffrey T. *America's Space Sentinels: DSP Satellites and National Security.* Lawrence: University Press of Kansas, 1999.

Richelson, Jeffrey T. *The Wizards of Langley: Inside the CIA's Directorate of Science and Technology.* Boulder, CO: Westview Press, 2002.

Rostow, Walt W. *Open Skies: Eisenhower's Proposal of July 21, 1955.* Austin: University of Texas Press, 1982.

Rothstein, Stephen M. "Ideas as Institutions: Explaining the Air Force's Struggle with its Aerospace Concept." PhD diss., Fletcher School of Law and Diplomacy, Tufts University, April 2006.

Ruffner, Kevin C., ed. *Corona: America's First Satellite Program.* CIA Cold War Records Series. Washington, DC: History Staff, Center for the Study of Intelligence, CIA, 1995.

Schemmer, Benjamin F. *The Raid.* New York: Harper & Row, 1976.

Schichtle, Colonel Cass, USAF. *The National Space Program From the Fifties into the Eighties.* Washington, DC: National Defense University Press, 1983.

Schnapper, Morris B., ed. *New Frontiers of the Kennedy Administration: The Texts of the Task Force Reports Prepared for the President.* Washington, DC: Public Affairs Press, 1961.

Schoettle, Enid Curtis Bok. "The Establishment of NASA." In *Knowledge and Power: Essays on Science and Government.* ed. Sanford A. Lakoff, 162–270. New York: Free Press, 1966.

Schwiebert, Ernest G. *A History of the U.S. Air Force Ballistic Missiles.* New York: Praeger, 1965.

Smith, M.V. *Ten Propositions Regarding Spacepower.* Maxwell AFB, AL: Air University Press, 2002.

Spires, David N. *Beyond Horizons: A Half Century of Air Force Space Leadership.* Revised Ed. Maxwell AFB, AL: Air University Press, 1998. Second edition titled *Beyond Horizons: A History of the Air Force in Space, 1947–2007*. Colorado Springs, CO: Air Force Space Command, Office of History, 2007.

Spires, David N., ed. Orbital Futures: Selected Documents in Air Force Space History. 2 vols. Peterson AFB, CO: Air Force Space Command Office of History, 2004.

Stares, Paul B. *The Militarization of Space: U.S. Policy, 1945–1984*. Ithaca, NY: Cornell University Press, 1985.

Stares, Paul B. *Space and National Security*. Washington, DC: Brookings Institution, 1987.

Steinberg, Gerald M. *Satellite Reconnaissance: The Role of Informal Bargaining*. New York: Praeger Publishers, 1983.

Strode, Rebecca V. "Commentary on the Soviet Draft Space Treaty of 1981." In *American Military Space Policy: Information Systems, Weapons Systems, and Arms Control*, ed. Colin S. Gray, 85–91. Cambridge, MA: Abt Books, 1982.

Taubman, Philip. *Secret Empire: Eisenhower, the CIA, and the Hidden Story of America's Space Espionage*. New York: Simon & Schuster, 2003.

Temple, L. Parker. *Shades of Gray: National Security and the Evolution of Space Reconnaissance*. Reston, VA: AIAA, 2005.

Terrell, Delbert R. Jr., (Col., USAFR). *The Air Force Role in Developing International Outer Space Law*. Maxwell AFB, AL: Air University Press, May 1999.

Toomay, John C. "Warning and Assessment Sensors." In *Managing Nuclear Operations*, eds. Ashton B. Carter, John D. Steinbruner, and Charles A. Zraket, 282–321. Washington, DC: Brookings Institution, 1987.

Trento, Joseph J. *Prescription for Disaster*. New York: Crown Publishers, 1987.

U.S. Military Uses of Space, 1945–1991: Guide and Index. Washington, DC: The National Security Archive and Chadwyck-Healey, Inc., Alexandria, VA, 1991.

Von Bencke, Matthew J. *The Politics of Space: A History of U.S.-Soviet/Russian Competition and Cooperation in Space*. Boulder, CO: Westview, 1997.

Watts, Barry D. *The Military Use of Space: A Diagnostic Assessment* Washington, DC: Center for Strategic and Budgetary Assessments, February 2001.

Wolfe, Tom. *The Right Stuff*. New York: Bantam Press, 1980.

Woodward, Bob. *Veil: The Secret Wars of the CIA, 1981–1987*. New York: Pocket Books, 1987.

Wertheimer, John. "The Antisatellite Negotiations." In *Superpower Arms Control: Setting the Record Straight*, eds. Albert Carnesale, and Richard N. Haass, 139–164. Cambridge, MA: Ballinger Publishers, 1987.

York, Herbert F. *Race to Oblivion: A Participant's View of the Arms Race*. New York: Simon & Schuster, 1970.

Journal Articles

Baucom, Donald R. "Space and Missile Defense." *Joint Force Quarterly* 33 (Winter 2002–2003): 50–55.

Beresford, Spencer M. "Preface to Naval Strategy in Outer Space." *U.S. Naval Institute Proceedings* 87 (March 1961): 33–41.

Berkowitz, Marc J. "Antisatellites and Strategic Stability." *Airpower Journal* 3 (Winter 1989): 46–59.

Bowman, Dr. Robert M. "Arms Control in Space: Preserving Critical Strategic Space Systems Without Weapons in Space." *Air University Review* 37 (November–December 1985): 58–72.

Brandt, Major General Thomas C. "Military Uses for Space." *Air University Review* 37 (November–December 1985): 40–51.

Brown, Colonel William W. "The Balance of Power in Outer Space." *Parameters* 7, no. 3 (1977): 8–15.

Cagle, Captain Malcolm W. USN "The Navy's future Role in Space." *U.S. Naval Institute Proceedings* 89 (January 1963): 86–93.

Carter, Ashton B. "Satellites and Anti-Satellites: The Limits of the Possible." *International Security* 10 (Spring 1986): 46–98.

Chayes, Abram, and Antonia Handler Chayes. "Testing and Development of 'Exotic' Systems under the ABM Treaty: The Great

Reinterpretation Debate." *Harvard Law Review* 99 (June 1986): 1956–1971.

Clark, Mark T. "The ABM Treaty Interpretation Dispute: Partial Analyses and the Forgotten Context." *Global Affairs* 2 (Summer 1987): 58–79.

Cooper, Henry F. "Anti-Satellite Systems and Arms Control: Lessons From the Past." *Strategic Review* 17 (Spring 1989): 40–48.

Coulter, Colonel John M. and Major Benjamin J. Loret. "Manned Orbiting Space Stations." *Air University Review* 16 (May–June 1965): 33–41.

Davis, Philip O., and William G. Holder. "Keynote of the 1970s: Joint Ventures into Space." *Air University Review* 23 (September–October 1973): 16–29.

DeBlois, Bruce M. "Space Sanctuary: A Viable National Strategy," *Airpower Journal* 12 (Fall 1998): 41–57.

Drew, Lieutenant Colonel Dennis M. "Of Leaves and Trees: A New View of Doctrine." *Air University Review* 33 (January–February 1982): 40–48.

Ehrhart, Major Robert C. "Some Thoughts on Air Force Doctrine." *Air University Review* 31 (March–April 1980): 29–38.

Ferguson, Major General James. "Manned Craft and the Ballistic Missile." *Air University Quarterly Review* 12 (Winter and Spring 1960–1961): 251–256.

Friedenstein, Lieutenant Colonel Charles D. "The Uniqueness of Space Doctrine." *Air University Review* 37 (November–December 1985): 13–23.

Gabriel, General Charles A. "The Air Force: Where We Are and Where We're Heading." *Air University Review* 35 (January–February 1984): 2–10.

Garfinkle, Adam M. "ABM—The Wrong Debate." *The National Interest*, Spring 1988, 76–84.

Garthoff, Raymond L. "Banning the Bomb in Outer Space." *International Security* 5 (Winter 1980/81): 25–40.

Graham, Lieutenant General Daniel O., USA, (Ret.). "Toward a New Strategy: Bold Strokes Rather Than Increments." *Strategic Review* 9 (Spring 1981): 9–16.

Gray, Dr. Colin S. "Space Arms Control: A Skeptical View." *Air University Review* 37 (November–December 1985): 73–86.

Hafner, Donald L. "Averting a Brobdingnagian Skeet Shoot: Arms Control Measures for Anti- Satellite Weapons." *International Security* 5 (Winter 1980/81): 41–60.

Hall, R. Cargill. "The Evolution of U.S. National Security Space Policy and Its Legal Foundations in the 20th Century." *Journal of Space Law* 33 (Summer 2007): 1–104.

Hall, R. Cargill. "Sputnik, Eisenhower, and the Formation of the U.S. Space Program." *Quest* 14 (2007): 32–39.

Halperin, Morton H. "The Gaither Report and the Policy Process." *World Politics* 13 (April 1961): 360–384.

Hansen, Lieutenant Colonel Richard Earl, USAF, (Ret.) "Freedom of Passage on the High Seas of Space." *Strategic Review* 5 (Fall 1977): 84–92.

Hays, Peter, and Karl Mueller, "Going Boldly—Where? Aerospace Integration, the Space Commission, and the Air Force's Vision for Space," *Aerospace Power Journal* 15, no. 1 (Spring 2001): 34–49.

Heftel, The Honorable Cecil. "A Space Policy for the 1980s—And Beyond." *Air University Review* 32 (November–December 1980): 2–16.

Henry, Major General R. C., and Major Aubrey B. Sloan. "The Space Shuttle and Vandenberg Air Force Base." *Air University Review* 27 (September–October 1976): 19–26.

Henry, Major Richard C. "The Immediate Mission in Space." *Air University Quarterly Review* 13 (Fall 1961): 30–44.

Herres, General Robert T. "The Future of Military Space Forces." *Air University Review* 38 (January–March 1987): 40–47.

Herres, General Robert T. "The Military in Space: A Historical Relationship." *Space Policy* 3 (May 1987): 92–95.

Holley, Major General I.B., Jr., USAF Reserve, (Ret.). "The Doctrinal Process: Some Suggested Steps." *Military Review* 59 (April 1979): 2–13.

Holley, Major General I.B., Jr., USAF Reserve, (Ret.). "Of Saber Charges, Escort Fighters, and Spacecraft: The Search for Doctrine." *Air University Review* 34 (September–October 1983): 2–11.

Hopwood, Major General Lloyd P. "The Military Impact of Space Operations." *Air University Quarterly Review* 10 (Summer 1958): 142–146.

Humble, Ronald D. "Space Warfare in Perspective." *Air University Review* 33 (July–August 1982): 81–86.

Jennings, Lieutenant Colonel Frank W., USAF Reserve, (Ret.) "Doctrinal Conflict Over the Word Aerospace." *Airpower Journal* 4 (Fall 1990): 46–58.

Johnson, Major General Oris B. "Space: Today's Front Line of Defense." *Air University Review* 20 (November–December 1968): 95–102.

Kane, Colonel Francis X. "Anti-Satellite Systems and U.S. Options." *Strategic Review* 10 (Winter 1982): 56–64.

Kane, Colonel Francis X. "The NASA Program." *Air University Quarterly Review* 14 (Winter and Spring 1962–1963): 189–204.

Keese, Major General William B. "Tomorrow's Role in Aerospace." *Air University Quarterly Review* 14 (Winter and Spring 1962–1963): 258–265.

Kraemer, Sven F. "The Krasnoyarsk Saga." *Strategic Review* 18 (Winter 1990): 25–38.

Krieger, Colonel Clifford R. "USAF Doctrine: An Enduring Challenge." *Air University Review* 36 (September–October 1984): 16–25.

Licklider, Roy E. "The Missile Gap Controversy." *Political Science Quarterly* 4 (December 1970): 600–615.

Logsdon, John M. "The Decision to Develop the Space Shuttle." *Space Policy* 2 (May 1986): 103–119.

Lorenzini, Lieutenant Colonel Dino A. "Space Power Doctrine." *Air University Review* 33 (July–August 1982): 16–21.

Lorenzini, Lieutenant Colonel Dino A., and Major Charles L. Fox. "2001: A U.S. Space Force." *Naval War College Review* 34 (March–April 1981): 48–67.

Lupton, Lieutenant Colonel David E., USAF, (Ret.) "Space Doctrines." *Strategic Review* 11 (Fall 1983): 36–47.

MacGregor, Lieutenant Colonel Charles H., and Major Lee H. Livingston. "Air Force Objectives in Space." *Air University Review* 29 (July–August 1978): 59–62.

McKee, Colonel Daniel D. "The Gemini Program." *Air University Review* 16 (May–June 1965): 6–15.

Myers, Colonel Kenneth A., USAF, and Lieutenant Colonel John G. Tockston, USAF. "Real Tenets of Military Space Doctrine." *Airpower Journal* 2 (Winter 1988): 54–68.

Ogle, Major General Dan C. "The Threshold of Space." *Air University Quarterly Review* 10 (Summer 1958): 2–6.

Pace, Scott. "US Space Transportation Policy: History and Issues for a New Administration." *Space Policy* 4 (November 1988): 307–318.

Parrington, Lieutenant Colonel Alan J., USAF. "US Space Doctrine: Time for a Change?" *Airpower Journal* 3 (Fall 1989): 51–61.

Piotrowski, General John L. "C³I for Space Control." *Signal* 41 (June 1987): 23–33.

Piotrowski, General John L. "A Soviet Space Strategy." *Strategic Review* 15 (Fall 1987): 55–62.

Piotrowski, General John L. "Space Based Wide Area Surveillance." *Signal* 43 (May 1989): 493–495.

Piotrowski, General John L. "U.S. Antisatellite Requirements: Myths and Facts." *Armed Forces Journal International* 125 (September 1987): 64–68.

Pipes, Richard. "Why the Soviet Union Thinks it Could Fight and Win a Nuclear War." *Commentary* 64 (July 1977): 134–146.

"Policy Focus: National Security and the U.S. Space Program After the Challenger Tragedy." *International Security* 11 (Spring 1987): 141–186.

Puckett, Dr. Robert H. "American Space Policy: Civilian/Military Dichotomy." *Air University Review* 16 (March–April 1965): 45–50.

Ritland, Major General Osmond J. "Space Systems." *Air University Quarterly Review* 14 (Winter and Spring 1962–1963): 177–188.

Roland, Alex. "Priorities in Space for the USA." *Space Policy* 3 (May 1987): 104–114.

Rosenberg, Major General Robert A. "The Air Force and Its Military Role in Space." *Air University Review* 37 (November–December 1985): 52–57.

Ruegg, Major General Robert C. "Aeronautical Systems." *Air University Quarterly Review* 14 (Winter and Spring 1962–1963): 145–158.

Sandborn, Colonel Morgan W. "National Military Space Doctrine." *Air University Review* 28 (January–February 1977): 75–79.

Schofield, Colonel Martin B. "Control of Outer Space." *Air University Quarterly Review* 10 (Summer 1958): 93–104.

Schriever, General Bernard A. "The Space Challenge." *Air University Review* 16 (May–June 1965): 3–4.

Schwetje, Kenneth F., and Donald E. Walsh. "Hypersonic Flight: The Need for a New Legal Regime." *IEEE AES Magazine* 4 (May 1989): 32–36.

Singer, Lieutenant Colonel S.E. "The Military Potential of the Moon." *Air University Quarterly Review* 11 (Summer 1959): 31–53.

Smart, General Jacob E., USAF, (Ret.) "Strategic Implications of Space Activities." *Strategic Review* 2 (Fall 1974): 19–24.

Sofaer, Abraham D. "The ABM Treaty and the Strategic Defense Initiative." *Harvard Law Review* 99 (June 1986): 1972–1985.

Summers, Colonel Harry G. Jr. USA (Ret.). "Military Doctrine: Blueprint for Force Planning." *Strategic Review* 20 (Spring 1992): 9–22.

"The USAF Reports to Congress." *Air University Quarterly Review* 10 (Spring 1958): 30–60.

Weinberger, Caspar W. "The USA Should Begin to Deploy a Strategic Defense System." *Space Policy* 3 (May 1987): 96–99.

Wohlstetter, Albert J. "The Delicate Balance of Terror." *Foreign Affairs* 37 (January 1959): 211–234.

Worden, Simon P., and Bruce P. Jackson. "Space, Power, and Strategy." *The National Interest,* Fall 1988, 43–52.

Worthman, Colonel Paul E. "The Promise of Space." *Air University Review* 20 (January–February 1969): 120–127.

Periodicals

Abt, Clark C. "Space Denial: Costs and Consequences." *Air Force/Space Digest* 46 (March 1963): 45–55.

"Air Force Space Command." *Air Force Magazine* 70 (May 1987): 99–101.

"Air Force Space Command." *Air Force Magazine* 71 (May 1988): 101–105.

"Air Force Space Command." *Air Force Magazine* 72 (May 1989): 67–68.

Baucom, Donald R. "The Rise and Fall of Brilliant Pebbles." *Journal of Social, Political & Economic Studies* 29 (Summer 2004): 143–190.

Berry, F. Clifton Jr. "Space Is a Place: An Interview With Lt. Gen. Richard C. Henry, USAF." *Air Force Magazine* 65 (June 1982): 36–42.

Boutwell, Jeffrey, and F. A. Long. "The SDI and U.S. Security." *Daedalus* 114, no. 3 (1985): 315–329.

Buenneke, Richard H. Jr. "The Army and Navy in Space." *Air Force Magazine* 73 (August 1990): 36–39.

Bunday, McGeorge et al. "The President's Choice: Star Wars or Arms Control." *Foreign Affairs* 63, no. 2 (1984–85): 264–278.

Butz, J. S., Jr. "Aerospace Plane: Answer to Rocketing Costs." *Air Force/Space Digest* 45 (May 1962): 42–50.

Butz, J. S. Jr. "Building Blocks . . . But No Building." *Air Force/Space Digest* 46 (April 1963): 56–66.

Butz, J. S. Jr. "Crisis in the Space Program." *Air Force/Space Digest* 50 (October 1967): 84.

Butz, J. S. Jr. "MOL: The Technical Promise and Prospects." *Air Force/Space Digest* 48 (October 1965): 42–46.

Butz, J. S. Jr. "RX for Spaceborne Deterrence." *Air Force/Space Digest* 45 (April 1962): 48–68.

Butz, J. S. Jr. "Under the Spaceborne Eye: No Place to Hide." *Air Force/Space Digest* 50 (May 1967): 93–98.

Canan, James W. "Bold New Missions in Space." *Air Force Magazine* 67 (June 1984): 88–93.

Canan, James W. "Coming Back in Space." *Air Force Magazine* 70 (February 1987): 44–51.

Canan, James W. "High Space Heats Up." *Air Force Magazine* 68 (July 1985): 60–67.

Canan, James W. "Mastering the Transatmosphere." *Air Force Magazine* 69 (June 1986): 48–54.

Canan, James W. "Our Blind Spots in Space." *Air Force Magazine* 71 (February 1988): 44–51.

Canan, James W. "Recovery in Space." *Air Force Magazine* 71 (August 1988): 73.

Canan, James W. "Space Gets Down to Earth." *Air Force Magazine* 90 (August 1990): 30–34.

Canan, James W. "Space Plan 2000." *Air Force Magazine* 68 (July 1985): 68–73.

"Congress Boosts Space Laser Funding." *Aviation Week & Space Technology*, August 23, 1982, 16.

Covault, Craig. "Antisatellite Weapon Design Advances." *Aviation Week & Space Technology*, June 16, 1980, 243–247.

Covault, Craig. "Astronauts to Launch Early Warning Satellite, Assess Manned Reconnaissance From Space." *Aviation Week & Space Technology*, November 18, 1991, 65–69.

Covault, Craig. "New Defense Space Policy Supports Manned Flight Role." *Aviation Week & Space Technology*, December 8, 1986, 18–20.

Covault, Craig. "Next SDI Delta Launch to Carry Multiple Research Payload." *Aviation Week & Space Technology*, November 17, 1986, 20–21.

Covault, Craig. "Recon Satellites Lead Allied Intelligence Effort." *Aviation Week & Space Technology*, February 4, 1991, 25–26.

Covault, Craig. "U.S. Space Command Focuses on Strategic Control in Wartime." *Aviation Week & Space Technology*, March 30, 1987, 83–84.

Covault, Craig. "USAF Missile Warning Satellites Providing 90-sec. Scud Attack Alert." *Aviation Week & Space Technology*, January 21, 1991, 60–61.

"Defense Department Unveils $1.2-Billion Asat Restructuring Plan." *Aviation Week & Space Technology*, March 16, 1987, 19–21.

Ferguson, Lieutenant General James, USAF. "A Decade of Cooperation—The Military-NASA Interface." *Air Force/Space Digest* 51 (October 1968): 71–73.

Ferguson, Lieutenant General James, USAF. "Needed: Military 'Stick Time' in Space." *Air Force/Space Digest* 46 (April 1963): 46–54.

Foley, Theresa M. "Brilliant Pebbles Testing Proceeds at Rapid Pace." *Aviation Week & Space Technology*, November 14, 1988, 32–33.

Fulghum, David A. "Key Military Officials Criticize Intelligence Handling in Gulf War." *Aviation Week & Space Technology*, June 24, 1991, 83.

Garwin, Richard L. "Star Wars: Shield or Threat?" *Journal of International Affairs* 39, no. 1 (1985): 31–44.

Gertz, Bill. "The Secret Mission of NRO." *Air Force Magazine* 76 (June 1993): 60–63.

"Government's First Responsibility." *Air Force Magazine* 69 (November 1986): 6–9.

Guemer, Gary L. "What is *Pioof?" *Foreign Policy* 59 (1985): 73–84.

Hammer, Charles. "The Door SDI Won't Shut." *Washington Monthly* 19, no. 2 (1987): 21–24.

Hartmann, Frederick H. "Still a Defensible Idea." *Proceedings of the U.S. Naval Institute* 119, no. 1 (January 1993): 50–53.

Hartung, William. "Star Wars Pork Barrel." *Bulletin of the Atomic Scientists* 42, no. 1 (1986): 20–24.

Heppenheimer, Thomas A. "What Edward Teller Did." *American Heritage of Invention & Technology* 16, no. 3 (2002): 34–38.

Herkin, Gregg. "The Earthly Origins of Star Wars." *Bulletin of the Atomic Scientists* 43, no. 8 (1987): 20–28.

Holm, Hans-Henrik. "Star Wars." *Journal of Peace Research* 23, no. 1 (1986): 1–8.

Iafner, Donald L. "Assessing the President's Vision: The Fletcher. Miller, and Hoffman Panels." *Daedalus* 114, no. 2 (1985): 91–107.

Iannotta, Ben. "NASP Officials Flesh Out Lower-Cost Hyflite Option." *Space News*, June 14–20, 1993, 24.

Jones, Rodney W., and Steven A. Hildreth. "Star Wars: Down to Earth, or Gleam in the Sky?" *Washington Quarterly* 7, no. 4 (1984): 104–111.

Kolcum, Edward H. "Defense Moving to Exploit Space Shuttle." *Aviation Week & Space Technology*, May 10, 1982, 40–42.

Kramer, Peter. "Star Wars." *History Today* 49, no. 3 (1999): 41–47.

"Laser Applications in Space Emphasized." *Aviation Week & Space Technology*, July 28, 1980, 62.

"Laser Weaponry Technology Advances." *Aviation Week & Space Technology*, May 25, 1981, 65.

Latham, Donald. "The GPS War." *Space News*, July 12–18, 1993, 15.

Leavitt, William. "The Air Force Contribution to the NASA Effort." *Air Force/Space Digest* 46 (April 1963): 69–74.

Leavitt, William. "Getting MOL Off the Pad." *Air Force/Space Digest* 48 (July 1965): 65–67.

Leavitt, William. "Military Space: 1966 A Growing Maturity." *Air Force/Space Digest* 49 (May 1966): 82–87.

Leavitt, William. "The National Space Effort—VII: A New Look at the US Space Effort." *Air Force/Space Digest* 42 (September 1959): 77–89.

Leavitt, William. "USAF—The Momentous Quarter Century: The Air Force and Space." *Air Force/Space Digest* 53 (September 1970): 92–100.

LeMay, General Curtis E. "Aerospace Power Is Indivisible." *Air Force/Space Digest* 44 (November 1961): 66–68.

LeMay, General Curtis E. "Keeping Space Free." *Air Force/Space Digest* 46 (April 1963): 40–43.

Loosbrock, John F. "Space: Laboratory and Battlefield." *Air Force/Space Digest* 43 (December 1960): 37–39.

Lynch, David J. "Toward a New Launcher Lineup." *Air Force Magazine* 76 (January 1993): 48–51.

Mann, Paul. "Nunn Affirms 1972 ABM Pact, Finding Kinetic Tests Illegal." *Aviation Week & Space Technology*, March 16, 1987, 21–23.

"The Military Mission in Space." *Air Force/Space Digest* 43 (December 1960): 68–88.

"Milstar Launched." *Aviation Week & Space Technology*, February 14, 1994, 28.

Milton, General T.R., USAF, (Ret.) . "Viewpoint: How Secure Is Space?" *Air Force Magazine* 71 (December 1988): 97.

Milton, General T.R., USAF, (Ret.). "Viewpoint: A New High Ground." *Air Force Magazine* 68 (December 1985): 102.

Morrison, David C. "Beating the Space Crisis." *National Journal*, March 28, 1987, 756–761.

"NASA Budget History." *Aviation Week & Space Technology*, March 16, 1992, 123.

"The National Space Effort—IV: ARPA." *Air Force/Space Digest* 42 (May 1959): 61–64.

"The National Space Effort—V: The US Army." *Air Force/Space Digest* 42 (July 1959): 59–61.

"The National Space Effort—VI: The US Navy." *Air Force/Space Digest* 42 (August 1959): 69–71.

"National Test Facility Hosts Simulation, War-Gaming Exercises." *Aviation Week & Space Technology*, November 7, 1988, 42.

"New USAF Organization to Intensify Space Focus." *Aviation Week & Space Technology*, October 26, 1981, 25.

Nusbaumer, Michael R., Judith A. Dilrio, and Robert D. Bailer. "The Boycott of 'Star Wars' by Academic Scientists: The Relative Roles of Political and Technical Judgment." *Social Science Journal* 31, no. 4 (1994): 375–388.

Oberg, James E. "A Dozen Anti-ASAT Fallacies." *Air Force Magazine* 68 (July 1985): 79–81.

Pace, Scott. "GPS: Challenged by Success." *Space News*, August 30–September 5, 1993, 15.

"Pentagon Preparing to Restart Antisatellite Program in January." *Aviation Week & Space Technology*, November 14, 1988, 33–34.

"Pentagon Studying Laser Battle Stations in Space." *Aviation Week & Space Technology*, July 28, 1980, 57–58, 61.

"Personnel Sought to Develop Defense Payloads." *Aviation Week & Space Technology*, May 24, 1982, 54.

Power, General Thomas S. "Military Aspects of Manned Spaceflight." *Air Force/Space Digest* 46 (June 1963): 51–57.

Ritland, Major General Osmond J. USAF. "Space—Tomorrow's Battleground?" *Air Force/Space Digest* 43 (December 1960): 89–90.

Rivkin, David B. Jr. "SDI: Strategic Reality or Never-Never Land?" *Strategic Review* 15, no. 3 (1987): 43–54.

Robinson, Clarence J. Jr. "Army Pushes New Weapons Effort." *Aviation Week & Space Technology*, October 16, 1978.

Robinson, Clarence J. Jr. "Beam Weapons Technology Expanding." *Aviation Week & Space Technology*, May 25, 1981, 40–53.

Robinson, Clarence J. Jr. "ICBM Intercept in Boost Phase Pushed." *Aviation Week & Space Technology*, July 17, 1978, 47–50.

Robinson, Clarence J. Jr. "Laser Technology Demonstration Proposed" and "Advance Made on High-Energy Laser." *Aviation Week & Space Technology*, February 23, 1981, 16–19.

Robinson, Clarence J. Jr. "Layered Defense System Pushed to Protect ICBMs." *Aviation Week & Space Technology*, February 9, 1981, 83–86.

Robinson, Clarence J. Jr. "Missile Defense Gains Support." *Aviation Week & Space Technology*, October 22, 1979, 14–17.

Robinson, Clarence J. Jr. "Space-Based Laser Battle Station Seen." *Aviation Week & Space Technology*, December 8, 1980, 36–40.

Robinson, Clarence J. Jr. "Technology Program Spurs Missile Intercept Advances." *Aviation Week & Space Technology*, June 5, 1978, 108–111.

Robinson, Clarence J. Jr. "U.S. To Test ABM System with MX." *Aviation Week & Space Technology*, March 19, 1979, 23–26.

Rushing, Janice H. "Ronald Reagan's 'Star Wars' Address: Mythic Containment of Technical Reasoning." *Quarterly Journal of Speech* 72, no. 4 (1986): 415–433.

"SAMSOs 25th Anniversary." *Air Force Magazine* 62 (August 1979): 39–63.

Schlesinger, James R. "Rhetoric and Realities in the Star Wars Debate." *International Security* 10, no. 1 (1985): 3–12.

Schlitz, William P. "USAF's Investment in the National Space Transportation System." *Air Force Magazine* 65 (November 1982): 106–112.

Scholin, Allan R. "Cape Canaveral—From Matador to Dyna-Soar: USAF's Space-Age Veterans." *Air Force/Space Digest* 46 (April 1963): 76–82.

Schriever, Lieutenant General Bernard A. "Needed: Manned Operational Capability in Space." *Air Force/Space Digest* 44 (November 1961): 79–80.

Schriever, Lieutenant General Bernard A. "Organizing Our Military Space Effort." *Air Force/Space Digest* 42 (July 1959): 62–64.

Schroeer, Dietrich. "Technological Progress in the SDI Programme." *Survival* 32, no. 1 (1990): 47–64.

"Science Advisor Hits Directed-Energy Efforts." *Aviation Week & Space Technology*, July 27, 1981, 26.

Smith, Bruce A. "Air Defense Group Urges Early Launch of Teal Ruby." *Aviation Week & Space Technology*, December 1, 1986, 34–36.

Smith, Bruce A. "Military Space System Applications Increasing." *Aviation Week & Space Technology*, March 9, 1981, 83–87.

Smith, Bruce A. "USAF Officer Cites Need to Plan Orbital Strategy." *Aviation Week & Space Technology*, June 22, 1981, 104–105.

Smith, Bruce A. "Vandenberg Readied for Shuttle Launch." *Aviation Week & Space Technology*, December 7, 1981, 49–52.

Smith, Jeff. "Reagan, Star Wars, and American Culture." *Bulletin of the Atomic Scientists* 43, no. 1 (1987): 19–25.

"Soviets Stage Integrated Test of Weapons." *Aviation Week & Space Technology*, June 28, 1982, 21.

"Space and National Security: A Symposium." *Air Force/Space Digest* 44 (November 1961): 71–84.

"Space and National Security: Symposium at Las Vegas." *Air Force/Space Digest* 45 (November 1962): 62–82.

"Space Command: A Major Command." *Air Force Magazine* 66 (May 1983): 96–97.

"Space Command." *Air Force Magazine* 67 (May 1984): 112–113.

"Space Command." *Air Force Magazine* 68 (May 1985): 100–102.

"Space Command." *Air Force Magazine* 69 (May 1986): 98–100.

"Space Command." *Air Force Magazine* 70 (May 1987): 99–101.

"'Star Wars:' The Politics of Defense." *Partisan Review* 53, no. 4 (1986): 534–562.

Steinberg, Gerald M. "SDI and Organizational Politics of Military R&D." *Armed Forces & Society* 13, no. 4 (1987): 579–598.

Stillson, Albert C. "Space Control: How? and How Much?" *Air Force/Space Digest* 42 (May 1959): 65–69.

"Technology Eyed to Defend ICBMs, Spacecraft." *Aviation Week & Space Technology,* July 28, 1980, 40–41.

Tsipis, Kosta. "Why SDI Won't Fly." *Harper's Magazine* 284, no. 6 (1992): 21–23.

Ulsamer, Edgar. "Aldridge on the Issues." *Air Force Magazine* 69 (July 1986): 84–89.

Ulsamer, Edgar. "Assuring Access to Space." *Air Force Magazine* 67 (November 1984): 80–84.

Ulsamer, Edgar. "The Battle for SDI." *Air Force Magazine* 68 (February 1985): 44–53.

Ulsamer, Edgar. "Charting a Course for SDI." *Air Force Magazine* 67 (September 1984): 106–121.

Ulsamer, Edgar. "How Vulnerable are USAF Military Space Systems?" *Air Force Magazine* 55 (June 1972): 35–40.

Ulsamer, Edgar. "In Focus: The 'Aerospace Force' Controversy." *Air Force Magazine* 65 (August 1982): 12.

Ulsamer, Edgar. "Military Imperatives in Space." *Air Force Magazine* 68 (January 1985): 92–98.

Ulsamer, Edgar. "A More Liberal, Avant Garde R & D Program." *Air Force Magazine* 63 (June 1980): 42–45.

Ulsamer, Edgar. "The Question of Soviet Orbital Bombs." *Air Force Magazine* 55 (April 1972): 74–75.

Ulsamer, Edgar. "Slick 6." *Air Force Magazine* 68 (November 1985): 47–48.

Ulsamer, Edgar. "Space: The Fourth Dimension." *Air Force Magazine* 65 (November 1982): 102–104.

Ulsamer, Edgar. "Spacecom: Setting the Course for the Future." *Air Force Magazine* 65 (August 1982): 48–55.

Ulsamer, Edgar. "What's Up in Space." *Air Force Magazine* 69 (February 1986): 46–53.

"USAF to Push Asat Tests Despite Congressional Ban." *Aviation Week & Space Technology*, December 22, 1986, 27.

"USAF Weighs In-Orbit Storage of Satellites." *Aviation Week & Space Technology*, June 7, 1982, 19.

"USAF's Space Command to be Established Sept. 1." *Aviation Week & Space Technology*, June 28, 1982, 30–31.

"Washington Outlook: Deception Confirmed, Test Exonerated." *Aviation Week & Space Technology*, September 13, 1993, 19.

Welsh, Dr. Edward C. "Peaceful Purposes: Some Realistic Definitions." *Air Force/Space Digest* 44 (November 1961): 73–77.

Welsh, Dr. Edward C. "The US Military Space Program—Insufficiently Understood." *Air Force/Space Digest* 52 (July 1969): 61.

Whisenand, Major General James F., USAF. "Military Space Efforts: The Evolutionary Approach." *Air Force/Space Digest* 45 (May 1962): 55.

Witze, Claude. "Congress Takes a Second Look at Space and National Defense." *Air Force/Space Digest* 46 (April 1963): 86–92.

Witze, Claude. "How Our Space Policy Evolved." *Air Force/ Space Digest* 45 (April 1962): 83–92.

Witze, Claude. "Let's Get Operational in Space: Walter Dornberger—Space Pioneer and Visionary." *Air Force/Space Digest* 48 (October 1965): 80–88.

Wysocki, Joseph. "GPS and Selective Availability—The Military Perspective." *GPS World*, July/August 1991, 38–40.

Yonas, Gerold. "The Strategic Defense Initiative." *Daedalus* 114, no. 2 (1985): 73–90.

Zimmerman, Peter D. "Pork Bellies and S.D." *Foreign Policy* 63 (1986): 76–87.

Zuckert, Honorable Eugene M. "Space and the Cold War." *Air Force/Space Digest* 46 (April 1963): 35–39.

Congressional Hearings and Reports

U.S. Congress. House. Committee on Appropriations. Subcommittee on the Department of Defense. *Department of Defense Appropriations for 1984: Hearings before Subcommittee on Department of Defense, Part 8.* 98th Cong., 1st sess., 1983.

U.S. Congress. House. Committee on Armed Services. *Defense Department Authorization and Oversight for Fiscal Year 1985: Hearings before the Committee on Armed Services, Part 2.* 98th Cong., 2nd sess., 1984.

U.S. Congress. House. Committee on Armed Services. *Fiscal Years 1965–1969 Defense Program and Fiscal Year 1965 Defense Budget: Hearings before the Committee on Armed Services.* 88th Cong., 1st sess., 1964.

U.S. Congress. House. Committee on Armed Services. *Hearings on Military Posture, Fiscal Year 1966: Hearings before the Committee on Armed Services.* 89th Cong., 1st sess., 1965.

U.S. Congress. House. Committee on Armed Services. Investigations Subcommittee. *Hearing before the Investigations Subcommittee on H.R. 5130: Aerospace Force Act.* 97th Cong., 2nd sess., May 19, 1982.

U.S. Congress. House. Committee on Armed Services. *Defense Department Authorization and Oversight for Fiscal Year 1986: Hearings before the Committee on Armed Services, Part 1.* 99th Cong., 1st sess., 1985.

U.S. Congress. House. Committee on Government Operations. *Organization and Management of Missile Programs: Hearings before the Committee on Government Operations.* 86th Cong., 1st sess., 1959.

U.S. Congress. House. Committee on Science and Astronautics. *A Chronology of Missile and Astronautic Events.* 87th Cong., 1st sess., 1961.

U.S. Congress. House. Committee on Science and Astronautics. *Defense Space Interests: Hearings before the Committee on Science and Astronautics.* 87th Cong., 1st sess., 1961.

U.S. Congress. House. Committee on Science and Astronautics. *Message from the President of the United States Transmitting the*

First Annual Report on the Nation's Activities and Accomplishments in the Aeronautics and Space Fields. 86th Cong., 1st sess., February 2, 1959.

U.S. Congress. House. Committee on Science and Technology. Subcommittee on Space Science and Applications. *The Need For a Fifth Space Shuttle Orbiter: Hearing before the Subcommittee on Space Science and Applications.* 97th Cong., 2nd sess., June 15, 1982.

U.S. Congress. House. Committee on Science and Technology. Subcommittee on Space Science and Applications. *The Need for an Increased Space Shuttle Orbiter Fleet.* 97th Cong., 2nd sess., 1982, Committee Print Serial HH.

U.S. Congress. Office of Technology Assessment. *Anti-Satellite Weapons, Countermeasures, and Arms Control.* OTA-ISC-281. Washington, D.C.: GPO, September 1985.

U.S. Congress. Office of Technology Assessment. *Ballistic Missile Defense Technologies.* Washington, D.C.: GPO, September 1985.

U.S. Congress. Senate. Committee on Aeronautical and Space Sciences. Subcommittee on Governmental Organization for Space Activities. *Investigation of Governmental Organization for Space Activities: Hearings before the Subcommittee on Governmental Organization for Space Activities.* 86th Cong., 1st sess., 1959.

U.S. Congress. Senate. Committee on Appropriations. Subcommittee on Defense Appropriations. *Defense Appropriations for Fiscal Year 1985: Hearings before the Subcommittee on Defense Appropriations, Part 3.* 98th Cong., 2nd sess., 1984.

U.S. Congress. Senate. Committee on Appropriations. Subcommittee on Defense Appropriations. *Defense Appropriations for Fiscal Year 1988: Hearings before Subcommittee on Defense Appropriations, Part 3.* 100th Cong., 1st sess., 1987.

U.S. Congress. Senate. Committee on Armed Services. *Authorization for Military Procurement, Research and Development, Fiscal Year 1970: Hearings before the Committee on Armed Services.* 91st Cong., 1st sess., 1969.

U.S. Congress. Senate. Committee on Armed Services. *Defense Authorization for Fiscal Years 1988 and 1989: Hearings before the Committee on Armed Services.* 100th Cong., 1st sess., 1987.

U.S. Congress. Senate. Committee on Armed Services. Preparedness Investigating Subcommittee. *Inquiry into Satellite and Missile Programs: Hearings before the Preparedness Investigating Subcommittee.* 85th Cong., 1st and 2nd sess., 1957–1958.

U.S. Congress. Senate. Committee on Armed Services and Subcommittee on Department of Defense of the Committee on Appropriations. *Military Procurement Authorizations, Fiscal Year 1966: Hearings before the Committee on Armed Services and the Subcommittee on Department of Defense of the Committee on Appropriations.* 89th Cong., 1st sess., 1965.

U.S. Congress. Senate. Committee on Commerce, Science, and Transportation. Subcommittee *on Science, Technology, and Space. NASA Authorization for Fiscal Year 1982: Hearings before the Subcommittee on Science, Technology, and Space, Part 2.* 97th Cong., 1st sess., 1981.

U.S. Congress. Senate. Committee on Commerce, Science, and Transportation. Subcommittee on Science, Technology, and Space. *NASA Authorization for Fiscal Year 1983: Hearings before the Subcommittee on Science, Technology, and Space.* 97th Cong., 2nd sess., 1982.

U.S. Congress. Senate. Committee on Commerce, Science, and Transportation. Subcommittee on Science, Technology, and Space. *Vandenberg Space Shuttle Launch Complex: Hearing before the Subcommittee on Science, Technology, and Space.* 98th Cong., 1st sess., September 10, 1984.

U.S. Congress. Senate. Committee on Foreign Relations. *Controlling Space Weapons: Hearings before the Committee on Foreign Relations.* 98th Cong., 1st sess., 1983.

U.S. Congress. Senate. Committee on Foreign Relations. *Treaty on Outer Space: Hearings before the Committee on Foreign Relations.* 90th Cong, 1st sess., 1967.

U.S. Congress. Senate. Committee on Foreign Relations. Subcommittee on Arms Control, Oceans, International Operations and Environment. *Arms Control and the Militarization of Space: Hearing before the Subcommittee on Arms Control, Oceans, International Operations and Environment.* 97th Cong., 2nd sess., September 20, 1982.

U.S. Congress. Senate. Senator Sam Nunn discussing the role of Congress in the ABM Treaty interpretation dispute. *Congressional Record*. Daily ed. (March 11–13, 1987), S2967–S2986, S3090–S3095, and S3171–S3173.

U.S. Congress. Senate. Subcommittee of the Committee on Appropriations. *Air Force National Programs: Special Hearing before a Subcommittee of the Committee on Appropriations.* 97th Cong., 2nd sess., 1982.

U.S. Congress. Senate. Subcommittee of the Committee on Appropriations. *Defense Appropriations for Fiscal Year 1988: Hearings before A Subcommittee of the Committee on Appropriations.* 100th Cong., 1st sess., 1987.

Government Documents, Studies, and Technical Reports

Aldridge, E. C. "Pete" Jr. "Assured Access: 'The Bureaucratic Space War.'" Dr. Robert H. Goddard Historical Essay. n.d. Offprint provided to author by the Office of the Secretary of the Air Force.

Aldridge, E. C. Jr. "MEMORANDUM FOR SECRETARY OF DEFENSE, SUBJECT: The Air Force Role in Space—INFORMATION MEMORANDUM." Department of the Air Force. Office of the Secretary, December 7, 1988.

America Plans for Space: A Reader Based on the National Defense University Space Symposium. Washington, DC: National Defense University Press, June 1986.

Boehm, Joshua et al. "A History of United States National Security Space Management and Organization." Space Commission Staff Background Paper. Washington, DC: Commission to Assess National Security Space Management and Organization, 2001.

Commission on Integrated Long-Term Strategy. Working Group on Technology. *Recommended Changes in U.S. Military Space Policies and Programs.* Washington, DC: Department of Defense, October 1988.

Department of the Air Force. *Air Force Manual 1–6: Aerospace Doctrine: Military Space Doctrine.* Draft, undated; OPR: XOXID [Colonel D.R. McNabb]; approved by Major General E.N. Block Jr. Copy provided to author by Dana J. Johnson, October 25, 1993.

Department of the Army. HQ USA. *Field Manual 100–5: Operations.* Washington, DC: GPO, May 5, 1986.

Downey, Arthur J. *The Emerging Role of the US Army in Space.* Washington, DC: National Security Affairs Monograph, National Defense University, 1985.

Holley, I.B. Jr. "An Enduring Challenge: The Problem of Air Force Doctrine." The Harmon Memorial Lecture Series in Military History. No. 16. Colorado Springs, CO: USAF Academy, 1974.

Holley, I.B. Jr. *Ideas and Weapons: Exploitation of the Aerial Weapon by the United States during World War I: A Study in the Relationship between Technological Advance, Military Doctrine, and Development of Weapons.* New Haven, CT: Yale University Press, 1953; Reprint, Washington, DC: Office of Air Force History, 1983.

HQ USAF. *Air Force Manual 1–1: Basic Aerospace Doctrine of the United States Air Force.* Vol. I and II. Washington, DC: GPO, March 1992.

HQ USAF. *Air Force Manual 1–6: Military Space Doctrine.* Washington, DC: GPO, October 15, 1982.

HQ USAF. Office of the Secretary of the Air Force. "Air Force Policy Letter for Commanders: SDI, The Cornerstone of Peacekeeping." October 1, 1987.

Johnson, Nicholas L. *The Soviet Year in Space, 1982.* Colorado Springs, CO: Teledyne Brown Engineering, January 1983; and subsequent versions of this annual report published in 1984–1990.

Joint Chiefs of Staff. *JCS Pub. 1: Department of Defense Dictionary of Military and Associated Terms.* Washington, DC: Joint Chiefs of Staff, April 1, 1984.

Joint Chiefs of Staff. *JCS Publication 2: Unified Action Armed Forces (UAAF).* Washington, DC: Joint Chiefs of Staff, December 1, 1986.

Kecskemeti, Paul. *The Satellite Rocket Vehicle: Political and Psychological Problems.* RAND RM-567. Santa Monica, CA: RAND Corporation, October 4, 1950.

Mitchell, Eddie. *Apogee, Perigee, and Recovery: Chronology of Army Exploitation of Space.* RAND Note N-3103-A. Santa Monica, CA: RAND Corporation, 1991.

Myers, Grover E. *Aerospace Power: The Case for Indivisible Application.* Maxwell AFB, AL: Air University Press, September 1986.

National Aeronautics and Space Administration. *Aeronautics and Space Report of the President, 1973 Activities.* Washington, DC: GPO, 1974; and subsequent editions of this annual report 1974–1990.

Piotrowski, General John L. "Military Space Imperatives." Address at Air Force Association Space Symposium. Colorado Springs, CO, May 21, 1987.

Ravenstein, Charles A. *The Organization and Lineage of the United States Air Force.* Washington, DC: Office of Air Force History, 1986.

Swan, Major Peter A., ed. *The Great Frontier—Military Space Doctrine: A Book of Readings for the United States Air Force Academy Military Space Doctrine Symposium, 1–3 April 1981.* Vol. I–IV. Colorado Springs, CO: USAF Academy, 1981.

U.S. Department of Defense. *Anti-Satellite: A Report to Congress.* Washington, DC, February 1990.

U.S. Department of Defense. "Department of Defense Space Policy (Unclassified)." Washington, DC, March 10, 1987.

U.S. Department of Defense. Directive 5160.32. "Development of Space Systems." Washington, DC, September 8, 1970.

U.S. Department of Defense. "Fact Sheet: DOD Space Policy." Washington, DC, August 11, 1982.

U.S. Department of Defense. Office of Assistant Secretary of Defense (Public Affairs). "News Release: SDI Gains Milestone I Approval." Washington, DC, September 18, 1987.

U.S. Department of Defense. *Soviet Military Power.* Washington, DC: GPO, September 1981; and subsequent editions of this annual report printed in March 1983, April 1984, April 1985, March 1986, March 1987, April 1988, and September 1989.

U.S. Department of Defense. Strategic Defense Initiative Organization. *Report to the Congress on the Strategic Defense Initiative.* Washington, DC: GPO, 1985; and subsequent editions of this annual report printed in 1986, April 1987, April 1988, March 13, 1989, and May 1990.

U.S. Department of Defense, and U.S. Department of State. *Soviet Strategic Defense Programs.* Washington, DC: GPO, October 1985.

U.S. Department of State. Bureau of Public Affairs. Statement of Ambassador Paul H. Nitze. "On the Road to a More Stable Peace." Current Policy No. 657, February 20, 1985.

U.S. Department of State. Bureau of Public Affairs. Statements by Special Advisor to the President and Secretary of State on Arms Control Matters Ambassador Paul H. Nitze and Legal Advisor Abraham D. Sofaer. "The ABM Treaty and the SDI Program." Current Policy No. 755, October 1985.

U.S. Department of State. *Foreign Relations of the United States, 1955–1957, Vol. XI: United Nations and General International Matters.* Washington, DC: GPO, 1990.

U.S. Department of State. *Foreign Relations of the United States, 1955–1957, Vol. XIX: National Security Policy.* Washington, DC: GPO, 1990.

U.S. Department of State. *Foreign Relations of the United States, 1958–1960, Vol. II: United Nations and General International Matters.* Washington, DC: GPO, 1991.

United States Space Foundation. *Fourth National Space Symposium Proceedings Report: Space Challenge '88.* Colorado Springs, CO: United States Space Foundation, 1988.

Viotti, Major Paul, ed. *Military Space Doctrine—The Great Frontier: Final Report for the United States Air Force Academy Military Space Doctrine Symposium, 1–3 April 1981.* Colorado Springs, CO: USAF Academy, 1981.

Welch, General Larry D., and Secretary E. C. Aldridge Jr. "MEM-ORANDUM FOR ALMAJCOM-SOA, SUBJECT: Air Force Space Policy—INFORMATION MEMORANDUM." Department of the Air Force. HQ USAF, December 2, 1988.

The White House. Executive Office of the President. National Aeronautics and Space Council. *Aeronautics and Space Report of the President, 1959 Activities.* February 22, 1960; and subsequent editions of this annual report released on January 18, 1961, January 27, 1964, January 27, 1965, January 31, 1966, January 31, 1967, January 30, 1968, January 17, 1969, January 1970, and March 1972.

The White House. *National Security Strategy of the United States.* Washington, DC: GPO, January 1993.

The White House. Office of the Press Secretary. "Fact Sheet: U.S. National Space Policy." November 16, 1989.

Documents from NSC Box at National Archives, Washington, D.C.

(Arranged Chronologically)

Rockefeller, Nelson A. "Memorandum for Mr. James S. Lay, Jr., Executive Secretary, National Security Council, Subject: U.S. Scientific Satellite Program." May 17, 1955.

National Security Council. NSC 5520. "U.S. Scientific Satellite Program." May 20, 1955.

U.S. Department of Defense. "Progress Report on the U.S. Scientific Satellite Program." October 3, 1956.

Lay, James S. Jr. "Memorandum for the National Security Council, Subject: Implications of the Soviet Earth Satellite for U.S. Security." October 11, 1957.

National Security Council. Planning Board. NSC 5814. "U.S. Policy on Outer Space (Draft)." June 20, 1958.

National Security Council. National Aeronautics and Space Council. NSC 5918. "U.S. Policy on Outer Space (Draft)." December 17, 1959.

National Security Council. Planning Board. NSC 6021. "Missiles and Military Space Programs (Draft)." December 14, 1960.

National Security Council. NSC 6108. "Certain Aspects of Missile and Space Programs." January 18, 1961.

Bundy, McGeorge. National Security Action Memorandum No. 129. "U.S.-U.S.S.R. Cooperation in the Exploration of Space." February 23, 1962; revised February 27, 1962.

Bundy, McGeorge. National Security Action Memorandum No. 144. "Assignment of Highest National Priority to the APOLLO Manned Lunar Landing Program." April 11, 1962.

Kennedy, John F. National Security Action Memorandum 156. May 26, 1962.

Bundy, McGeorge. NSAM 183. August 27, 1962.

Kaysen, Carl. National Security Action Memorandum No. 191. "Assignment of Highest National Priority to Project DEFENDER." October 1, 1962.

Kaysen, Carl. National Security Action Memorandum No. 192. October 2, 1962.

Bundy, McGeorge. National Security Action Memorandum No. 237. "Project MERCURY Manned Space Flight (MA-9)." May 3, 1963.

Kennedy, John F. National Security Action Memorandum No. 271. "Cooperation with the USSR on Outer Space Matters." November 12, 1963.

Johnson, Lyndon B. National Security Action Memorandum No. 285. "Cooperation with the USSR on Outer Space Matters." March 3, 1964.

Bundy, McGeorge. National Security Action Memorandum No. 50. "Official Announcements of Launching into Space of Systems Involving Nuclear Power In Any Form." May 12, 1961; revision April 10, 1965.

Rostow, W.W. National Security Action Memorandum No. 354. "U.S. Cooperation with the European Launcher Development Organization (ELDO)." July 29, 1966.

Kissinger, Henry A. National Security Decision Memorandum 187. "International Space Cooperation—Technology and Launch Assistance." August 30, 1972.

Brzezinski, Zbigniew. Presidential Directive/NSC-25. "Scientific or Technological Experiments with Possible Large-Scale Adverse Environmental Effects and Launch of Nuclear Systems into Space (C)." December 14, 1977.

Carter, Jimmy. Presidential Directive/NSC-37. "National Space Policy (U)." May 11, 1978.

Brzezinski, Zbigniew. Presidential Directive/NSC-42. "Civil and Further National Space Policy (U)." October 10, 1978.

Brzezinski, Zbigniew. Presidential Directive/NSC-54. "Civil Operational Remote Sensing (U)." November 16, 1979.

Carter, Jimmy. Presidential Directive/NSC-59. "Nuclear Weapons Employment Policy (C)." July 25, 1980.

National Security Council. National Security Decision Directive Number 42. "National Space Policy (U)." July 4, 1982.

National Security Council. National Security Decision Directive 50. "Space Assistance and Cooperation Policy." August 6, 1982.

Clark, William P. NSDD 80. "Shuttle Orbiter Production Capability." February 3, 1983.

National Security Council. NSDD 85. "Eliminating the Threat From Ballistic Missiles." March 25, 1983.

National Security Council. NSDD 94. "Commercialization of Expendable Launch Vehicles." May 16, 1983.

National Security Council. NSSD 03-85. "Shuttle Pricing." January 24, 1985.

National Security Council. NSDD 164. "National Security Launch Strategy." February 25, 1985.

National Security Council. NSDD 181. "Shuttle Pricing for Foreign and Commercial Users." July 30, 1985.

National Security Council. NSSD 06-85. "National Space Transportation and Support Study." May 14, 1985.

National Security Council. National Security Decision Directive Number 172. "Presenting the Strategic Defense Initiative." May 30, 1985.

National Security Council. NSDD 254. "United States Space Launch Strategy." December 27, 1986.

Newspaper Articles

Abramson, Rudy. "New Space Satellite-Killer Tests Planned." *Los Angeles Times*. March 11, 1987, p. 11.

"Anti-Satellite System Tested Successfully." *Los Angeles Times*. August 23, 1986, p. 24.

Broad, William J. "Military Launches First New Rocket for Orbital Loads." *New York Times*. September 6, 1988, p. 1.

Broad, William J. "Pentagon Leaving Shuttle Program." *New York Times*. August 7, 1989, p. A13.

Carrington, Tim, and John Wolcott. "Gaping Gap: Pentagon Lags in Race to Match the Soviets in Rocket Launchers." *Wall Street Journal*. July 12, 1988, p. 1.

Friedman, Thomas L. "U.S. Formally Rejects 'Star Wars' in ABM Treaty." *New York Times*. July 15, 1993, p. A6.

Gordon, Michael R. "Air Force to Test a Weapon in Space." *New York Times*. February 20, 1986, p. A18.

Gordon, Michael R. "General Quitting as Project Chief for Missile Shield." *New York Times*. September 28, 1988, p. 1.

"The Kremlin Apology: Excepts From Speech." *New York Times*. October 25, 1989, p. A18.

Oberg, James E. "A Mysterious Soviet Space Launch." *Wall Street Journal*. January 21, 1986, p. 16.

Rosenthal, Andrew. "Pentagon: The New 'Star Wars' Chief Brings a Soft-Sell Approach to his Mission." *New York Times*. May 18, 1989, p. A15.

Rusher, Bill. "Why Brass Fights SDI." *Colorado Springs Gazette Telegraph.* August 13, 1989, p. 11.

Schmitt, Eric. "Spy-Satellite Unit Faces a New Life in Daylight." *New York Times.* November 3, 1992, p. A16.

Weiner, Tim. "Lies and Rigged 'Star Wars' Test Fooled the Kremlin, and Congress." *New York Times.* August 18, 1993, p. A1.

Zelnick, Robert C. "Obituary for Star Wars." *New York Times.* October 26, 1988, p. A28.

Internet Sources

Air Force Space Command. "Air Force Space Command." www.afspc.af.mil (accessed July 17, 2010).

Arlington Cemetery. "Arlington Cemetery." www.arlingtoncemetery.net (accessed June 24, 2010).

Central Intelligence Agency. "Central Intelligence Agency." www.cia.gov (accessed July 17, 2010).

Defense Advanced Research Projects Agency. "Defense Advanced Research Projects Agency." www.darpa.mil (accessed July 17, 2010).

Department of Defense. "Department of Defense." www.defense.gov (accessed July 17, 2010).

Jet Propulsion Laboratory. "Jet Propulsion Laboratory." www.jpl.nasa.gov (accessed July 17, 2010).

Joint Chiefs of Staff. "Joint Chiefs of Staff." www.jcs.mil (accessed July 17, 2010).

National Aeronautics and Space Agency. "National Aeronautics and Space Agency." www.hq.nasa.gov (June 24, 2010).

National Oceanic and Atmospheric Agency. "National Oceanic and Atmospheric Agency." www.noaa.gov (accessed July 17, 2010).

National Reconnaissance Office. "National Reconnaissance Office." www.nro.gov (accessed July 17, 2010).

Naval Research Laboratory. "Naval Research Laboratory." www.nrl.navy.mil (accessed July 17, 2010).

Office of the Secretary of Defense. "Office of the Secretary of Defense." www.defense.gov/osd (accessed July 17, 2010).

Research and Development. "Research and Development." www.rand.org (accessed July 17, 2010).

The White House. "National Security Council." www.whitehouse.gov/administration/eop/nsc (accessed July 17, 2010).

The White House. "Office of Science and Technology Policy." www.whitehouse.gov/administration/eop/ostp (accessed July 17, 2010).

United States Strategic Command. "United States Strategic Command." www.stratcom.mil (accessed July 17, 2010).

Wikipedia. "Wikipedia: The Free Encyclopedia." en.wikipedia.org (accessed June 24, 2010).

List of Acronyms

ABMA Army Ballistic Missile Agency

ABMT Anti-Ballistic Missile Treaty

ACDA Arms Control and Disarmament Agency (State Department)

AFB Air Force Base

AFBMD Air Force Ballistic Missile Division

AFSATCOM Air Force Satellite Communications System

AFSC Air Force Systems Command

AFSCF Air Force Satellite Control Facility

AFSPC Air Force Space Command

ALMV Air-Launched Miniature Vehicle

ALS Advanced Launch System

ARDC Air Research and Development Command (USAF)

ARPA Advanced Research Projects Agency (DoD)

ASAT Anti-Satellite

ASD(C3I) Assistant Secretary of Defense for Command, Control, Communications, and Intelligence

ASD(NII) Assistant Secretary of Defense for Networks and Information Integration

ASTP Apollo-Soyuz Test Project

BE Brilliant Eyes

BMD Ballistic Missile Defense

BP Brilliant Pebbles

CIA Central Intelligence Agency

CJCS Chairman of the Joint Chiefs of Staff

CNO Chief of Naval Operations

COMINT   Communications Intelligence

COPOUS   Committee on the Peaceful Uses of Outer Space (United Nations)

DARPA   Defense Advanced Research Projects Agency (DoD)

DCI   Director of Central Intelligence

DDR&E   Director of Defense Research and Engineering (DoD)

DDS&T   Deputy Director for Science and Technology (CIA)

DEW   Directed Energy Weapon

DIA   Defense Intelligence Agency

DMSP   Defense Meteorological Satellite Program

DoD   Department of Defense

DSB   Defense Science Board (DoD)

DSCS   Defense Satellite Communications System

DSP   Defense Support Program

Dyna-Soar   Dynamic Soaring

ELINT   Electronic Intelligence

EORSAT   ELINT Ocean Reconnaissance Satellite (Soviet Union)

ETR   Eastern Test Range

FLTSATCOM   Fleet Satellite Communications System

FOBS   Fractional Orbital Bombardment System (Soviet Union)

FRUS   *Foreign Relations of the United States*

FY   Fiscal Year

GAO   Government Accountability Office

GEO   Geostationary-Earth Orbit

GPALS   Global Protection against Limited Strikes

GPS   Global Positioning System

HASC   House Armed Services Committee

HOE   Homing Overlay Experiment

HPSCI   House Permanent Select Committee on Intelligence

IA   Interim Agreement (SALT I)

ICBM   Intercontinental Ballistic Missile

IGY   International Geophysical Year

INF   Intermediate-Range Nuclear Forces

IOC   Initial Operational Capability

IRBM Intermediate-Range Ballistic Missile

JCS Joint Chiefs of Staff

JPL Jet Propulsion Laboratory

KEW Kinetic Energy Weapon

KH Keyhole (Spysat Camera System)

LCI Legally Correct Interpretation (ABM Treaty)

LEO Low-Earth Orbit

LPAR Large Phased-Array Radar

LTBT Limited Test Ban Treaty

MAD Mutual Assured Destruction

MFP Major Force Program (DoD)

MHV Miniature Homing Vehicle

MIDAS Missile Detection and Alarm System

MOL Manned Orbital Laboratory

MOU Memorandum of Understanding

MSI Multi-Spectral Imaging

NACA National Advisory Committee on Aeronautics

NASA National Aeronautics and Space Administration

NASC National Aeronautics and Space Council (White House)

NASP National Aerospace Plane

NCA National Command Authority

NIE National Intelligence Estimate

NORAD North American Air Defense Command

NRL Naval Research Laboratory

NRO National Reconnaissance Office

NSAM National Security Action Memorandum (Kennedy Administration)

NSC National Security Council (White House)

NSDD National Security Decision Directive (Reagan Administration)

NSDM National Security Decision Memorandum (Ford Administration)

NSpC National Space Council (White House)

NSSD National Security Study Directive (Reagan Administration)

OMB Office of Management and Budget (White House)

OPP Other Physical Principles (ABM Treaty)

OSD   Office of the Secretary of Defense

OSI   On-Site Inspection

OST   Outer Space Treaty

OSTP   Office of Science and Technology Policy (White House)

PD   Presidential Directive (Carter Administration)

PPD   Presidential Policy Directive (Obama Administration)

PRC   Policy Review Committee (Carter Administration)

PRM   Presidential Review Memorandum (Carter Administration)

PSAC   President's Science Advisory Committee (Eisenhower Administration)

RORSAT   Radar Ocean Reconnaissance Satellite (Soviet Union)

SAC   Strategic Air Command

SAINT   Satellite Inspector

SALT   Strategic Arms Limitation Treaty (and Talks)

SAMSO   Space and Missile Systems Organization (Air Force)

SASC   Senate Armed Services Committee

SBSS   Space-Based Surveillance System

SDI   Strategic Defense Initiative

SDIO   Strategic Defense Initiative Organization

SecAF   Secretary of the Air Force

SFRC   Senate Foreign Relations Committee

SIGINT   Signals Intelligence

SLC   Space Launch Complex

SMC   Space and Missile Systems Center (Air Force)

SSA   Space Situational Awareness

SSCI   Senate Select Committee on Intelligence

SSTS   Space Surveillance and Tracking System

STS   Space Transportation System

TCP   Technological Capabilities Panel (Eisenhower Administration)

TENCAP   Tactical Exploitation of National Space Capabilities

UDRE   Under Secretary of Defense for Research and Engineering

UN   United Nations

UNGA   United Nations General Assembly

USA   United States Army

USAF   United States Air Force

USAFA   United States Air Force Academy

USD(A)   Under Secretary of Defense for Acquisition

USD(AT&L)   Under Secretary of Defense for Acquisition, Technology, and Logistics

USD(I)   Under Secretary of Defense for Intelligence

USD(P)   Under Secretary of Defense for Policy

USecAF   Under Secretary of the Air Force

USN   United States Navy

USSPACECOM   United States Space Command

USSTRATCOM   United States Strategic Command

VAFB   Vandenberg Air Force Base

WDD   Western Development Division

Index

About the Author

Peter L. Hays, PhD, retired after 25 years of service as an Air Force officer and is a senior scientist for the Science Applications International Corporation supporting the National Security Space Office in the Pentagon by helping to develop and implement national space policies and strategies. He recently served as chief of staff for the National Defense University Spacepower Theory Study and is currently associate director for studies for the Eisenhower Center for Space and Defense Studies at the USAF Academy and professorial lecturer in International Affairs at George Washington University's Space Policy Institute. Previous work experiences include service as executive editor of *Joint Force Quarterly*, professor at the School of Advanced Airpower Studies, director of the USAF Institute for National Security Studies, division chief for International Relations and Defense Policy in the Department of Political Science at the USAF Academy, and internships at the White House Office of Science and Technology Policy and National Space Council. He holds a PhD in International Relations from the Fletcher School of Law and Diplomacy, an MA in Defense and Strategic Studies from the University of Southern California, and is a 1979 Honor Graduate of the USAF Academy. Major publications include *United States Military Space* (2002), "Going Boldly—Where?" (2001), *Spacepower for a New Millennium* (2000), *Countering the Proliferation and Use of Weapons of Mass Destruction* (1998), and *American Defense Policy* (1997).